Modern Electronic Maintenance Principles

by D.J. Garland and F.W. Stainer

AF854472

Pergamon Press Limited

OXFORD · NEW YORK · TORONTO
SYDNEY · BRAUNSCHWEIG

Pergamon Press Ltd., Headington Hill Hall, Oxford

Pergamon Press Inc., Maxwell House, Fairview Park, Elmsford, New York 10523

Pergamon of Canada Ltd., 207 Queen's Quay West, Toronto 1

Pergamon Press (Aust.) Pty. Ltd., 19a Boundary Street, Rushcutters Bay, N.S.W. 2011, Australia

Vieweg & Sohn GmbH, Burgplatz 1, Braunschweig

Printed in Great Britain

First Edition 1970

Library of Congress Catalog Card No. 74-101488

08 014188 9

CONTENTS

Page

THE AUTHORS

F.W. Stainer was, until 1967, a member of the Research Branch, Headquarters Technical Training Command, Royal Air Force. He was responsible for most of the RAF research on the subject matter of this book, including working out the new methods which are advocated, and carrying out experiments on technicians under training in order to test the methods he evolved. In 1967 he was loaned by the Ministry of Aviation to the European Space Research Organization, and is now manager of the ESRO satellite telemetry station in the Falkland Islands, where he is in charge of the operation and maintenance of the station. He is an Associate Member of the Institution of Electrical Engineers.

D.J. Garland, B.Sc., C.Eng., MIEE, MIERE, at present a Group Captain in the Royal Air Force, was associated for some years with the work at Headquarters Technical Training Command in his capacity as a staff officer responsible for the training of all electronic and guided missile maintenance technicians of the Royal Air Force. He was connected closely with the introduction of the new maintenance training methods described in this book, and has been concerned, as a training officer at various RAF radio schools, with the implementation of these methods. He is at present a staff officer at the RAF Group Headquarters which is responsible for the efficiency of all technician training in the Royal Air Force.

ACKNOWLEDGEMENTS

The authors would like to thank Mr. A.J. Kenward for helpful suggestions, and the British Motor Corporation, Ltd., for permission to reproduce Figure 8.1 (a).

Figures 8.1b, 8.2a, 8.2b and 9.5 are reproduced by permission of the Controller of Her Majesty's Stationery Office.

PREFACE

The ever-increasing complexity of electronic equipment continuously increases the demands made on the maintenance technician. Until recently reliance has been placed on theoretical training, leaving the techniques of maintenance to be gained by experience. This approach to training for maintenance results in the need to obtain technicians of high ability and to provide long and expensive training courses. It is unfortunately true that most courses for technicians contain little or no instruction in the principles of maintenance or in systematic, logical procedures supported by efficient and instructive aids to maintenance. The reason for this omission undoubtedly lies in the lack of suitable information.

The problem of training technicians to maintain modern, complex electronic equipment has faced the Royal Air Force acutely in recent years, and research investigations and experiments were conducted in order to find an economical and logical method of training for maintenance. These efforts resulted in the adoption of new methods of training which are based upon the principles described in this book.

To the best of the authors' knowledge this is the first book dealing with the logical bases of electronic maintenance which is suitable for technicians, but it should also interest engineers, technologists and supervisors who have an interest in, or responsibility for, maintenance or the training of maintenance personnel. The bias of the book is electronic, but the principles described are applicable to other complex systems such as instrumentation, hydraulics and electric power.

This is not a servicing handbook, but deals with the fundamentals of maintenance. It covers, with examples, the basic principles of maintenance using logical methods which have been developed recently. It describes new aids to maintenance which can assist the technician in his maintenance task. Unfamiliar subjects such as the half-split method of fault location, functional diagrams and fault finding guides are explained.

Because reliability is so closely connected with maintenance, and is so important to the maintenance technician, this subject is explained in some detail. The few mathematical proofs are confined to the Appendices to Chapters 5 and 6, and can be omitted. Otherwise, there is very little mathematics in the book, although an exponential expression appears in connection with reliability.

Use is made of the modern graphical symbols that were introduced in British Standard BS 3939, and the names of quantities and units conform to the SI system of units which is soon to be introduced generally into British industry. The note on page ix gives further details.

The book should interest technicians pursuing Ordinary and Higher National Certificate courses and City and Guilds technicians' courses ; students following these courses are expected to know the fundamentals of logical fault-finding. It should also interest engineers, training staffs and maintenance technicians everywhere.

ABBREVIATIONS AND SYMBOLS

ABBREVIATIONS

This country's impending change from its present system of units to the metric system coincides with the introduction of a set of internationally recognized units known as Système International (SI) units. As far as electrical units are concerned, SI units are almost the same as Metre-Kilogram-Second (MKS) units, and common units such as volts, amperes, farads, henrys and ohms remain unchanged.

The only recently-named SI unit that occurs in this book is hertz (Hz), the unit of frequency: one hertz is a frequency of one cycle per second. Multiples of hertz are formed in the usual way:

1000 Hz = 1 kHz (kilohertz)

1000 kHz = 1 MHz (megahertz)

1000 MHz = 1 GHz (gigahertz)

Logarithm to base 10 is written log, and logarithm to base e is ln.

SYMBOLS

The authority for symbols to be used in circuit diagrams has, for some years, been a British Standard, BS 530 : 1948 (Graphical Symbols for Telecommunications). This standard is being superseded by BS 3939 (Graphical Symbols for Electrical Power, Telecommunications and Electronics Diagrams), which, at the time of writing, is still being published in sections. The symbols used in this book are up-to-date at the time of going to press, but it may be that some symbols will be changed as more of those in BS 530 are superseded by BS 3939.

Some of the new symbols are shown below; the biggest change is undoubtedly in the new internationally agreed symbols for resistors.

Screened conductor 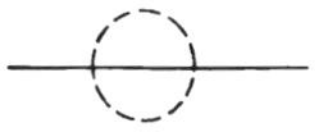

Fixed resistor 

Fixed resistor with fixed tapping (voltage divider) 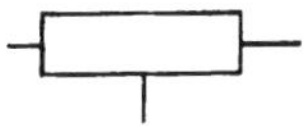

Fixed resistor with variable tapping (potentiometer) 

Variable resistor 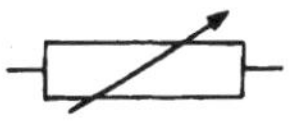

Plug

Socket 

Microphone 

Inductor

Switch 

CHAPTER 1 THE NEED FOR MAINTENANCE PRINCIPLES

THE PROBLEMS OF COMPLEXITY

The high cost of maintaining modern electronic equipment—retaining it in, or restoring it to, a serviceable state—presents a major problem to those who use the equipment. The problem is aggravated by the tendency to build and use equipment of ever-increasing complexity. It is a strange fact that the more reliable electronic components become, the greater is the problem of maintenance. The modern, more reliable components enable equipments to be designed which, because of the huge number of components that are incorporated, were just not feasible a few years ago. Because of the large number of components used, and the associated increase of interconnections between them, the circuits are complex. This complexity produces two disadvantageous effects.

The first is that although the reliability of an individual component may have improved with advancing technology, the number of components used is so large that the probability of **something** failing, and hence rendering the equipment unserviceable, is quite high during a fairly short period of time. The reliability of the system may therefore be no better than that of less complex systems using less reliable components. For example, a unit containing 10 components, each likely to fail twice in 100 hours of operation, could be expected to fail 20 times in that period. If the number of components is doubled, then the likelihood of failure of each component must be reduced to once in 100 hours if the unit is to maintain the same failure rate. It is much easier to double the number of components than to double their reliability!

The second effect is that due to the complexity of the equipment the technician will take longer to trace a fault than on simpler equipment. The reasons for this are that there are more possible sources of the fault, the relationships between the components are more difficult to deal with, and the technician has a much more extensive set of fault indications to analyze.

Computers provide a good illustration. Before the transistor was invented one of the biggest factors affecting the reliability of electronic equipment was the high failure rate of valves, and systems employing tens of thousands of valves were not feasible on the grounds of unreliability alone. Transistors are, however, very much more reliable than valves, and computers employing tens of thousands of transistors are feasible, and are made. When a fault does occur, however, it takes a long time to localize it. To avoid having the whole computer unserviceable for long periods, modern design is such that the computer is built of replaceable units, each consisting of a number of components, and the fault is first localized to a particular unit. This can then be changed rapidly, thus keeping the computer serviceable, and the faulty component within the unit can be traced and rectified separately. A price is paid for this reduction in the time that the computer is unserviceable, and it is that the user has to keep a stock of units as well as of components.

The computer is an example which illustrates the situation where a particular policy has been adopted with a view to making its operation and maintenance of maximum efficiency. This approach should be made to all equipment. The relative priority given to all the factors concerned in the maintenance of a particular equipment is referred to as the **maintenance policy** for that equipment. The factors which have to be considered in formulating a maintenance policy are discussed in Chapter 4.

MAINTENANCE TECHNICIANS

The requirements for a maintenance technician which are generally accepted at present are a comprehensive background of fundamental theory, basic manual skills such as the use of hand tools and soldering, some general guidance regarding fault location, such as the types of fault usually encountered in a particular equipment, and sufficient experience to develop the ability to trace faults quickly. Emphasis has always been placed on the theoretical aspect, and it is for this reason that its relevance to the maintenance task has been the subject of considerable research. Some of this research is discussed in Chapter 12.

It is not intended in this book to discuss how much theory is necessary for a technician, but only to note that as new electronic devices are discovered and applied the amount of supporting theory expands. Further, because the equipments being considered are late 20th-century, they are usually not easy to understand. This results in the need to recruit technicians of high intellectual calibre who are able to absorb the background knowledge, and in the need to provide long and expensive training courses.

Improved performance in the maintenance task can be expected if time is spent in analyzing the knowledge and skill required for the task, in developing aids to the task, and in working out methods of imparting the necessary knowledge and skills efficiently. New methods are, of course, only acceptable if they prove themselves in realistic situations. Work of this nature has been carried out in recent years, but the results are not generally available. It is undoubtedly for this reason that most existing courses for maintenance technicians include instruction neither in the principles of maintenance, nor in systematic, logical procedures supported by efficient and instructive maintenance aids. This book is intended to help to fill the gap.

A LOGICAL APPROACH

The problems of training technicians to maintain modern complex electronic equipment have been most acute in the armed Services, particularly where recruitment is voluntary and sufficient numbers of recruits of high ability are difficult to obtain. This has led to a lot of research into maintenance methods and training being undertaken, and some of this book stems from this research and the application of the results. Although the origins may be military, the results are directly applicable to any industry which is faced with the same problems.

Investigations have shown that the most important aspect of fault diagnosis is knowledge of the fundamental structure of equipment, i.e. knowledge of the relationships between the components of the equipment. It is possible to prepare diagrams which emphasize this information. These are termed **functional diagrams**, and they are described in detail in Chapter 8. It is also possible to prepare rules which define the best method of narrowing down the faulty area; these are based on the fundamental half-split rule which is described in Chapter 6. Having decided upon the information which is required for maintenance and the best method of displaying it, and having decided upon the best method of using this information, the following questions arise. How is this knowledge best conveyed to the technician? Should the maintenance be automated or should a man be trained to carry it out? These questions are also fully discussed in this book.

CHAPTER 2 RELIABILITY

INTRODUCTION

Maintenance cannot be divorced from reliability; if an equipment could be made perfectly reliable no maintenance would be necessary at all. It is essential therefore for the maintenance technician to know something about reliability and its relationship with the problems of maintenance. A knowledge of which components are most reliable, and which least reliable, assists in the maintenance task. The effects on reliability, and hence on maintenance, of such factors as temperature, humidity and vibration are also important, as are special methods, such as redundancy, by which reliability can be improved at the design stage.

This chapter discusses the connection between reliability and maintenance, and also defines the terms 'failure rate', 'reliability', 'mean time between failures', and 'availability'. The connection between these quantities is developed, and examples are given using typical component and equipment reliability figures. The method of calculating the reliability of a system is shown, major causes of unreliability are stated, and methods of improving reliability are described.

WHY PERFECT RELIABILITY CANNOT BE ACHIEVED

In theory maintenance could be avoided completely by making equipment perfectly reliable. In practice, of course, perfect reliability cannot be achieved. No-one has ever made anything that never wears out, never has a defect, and that would last for ever. Perfect reliability is like perpetual motion – interesting to imagine, but not practical.

But it may be thought that a piece of equipment could be made of components that were so reliable that maintenance would be unnecessary before it became out-of-date. Again, in theory this is possible, and, in fact, development of new components which are more reliable than the old is a ceaseless activity. However, we have already mentioned the fact that increases in component reliability usually lead to an increase in complexity. Also, the development of more reliable components costs money; research and testing have to be paid for. Production of more reliable components involves the introduction of new methods and machinery for manufacture, and of more stringent production testing.

A stage is reached where the cost of producing more reliable components outweighs the savings that are made by reduced maintenance costs and reduced holdings of spares. When this stage is reached, improving the reliability of the equipment becomes too expensive to be practical.

The nearest approach that has been made to eliminating maintenance completely is in the manufacture of unmanned satellites, for when they are in orbit maintenance is extremely difficult. It is undoubtedly due to the development of earth satellites that great interest has been taken in reliability in recent years. However, the astronomical cost of achieving the standard of reliability which is required is well known, as is the fact that failures still occur.

A point is therefore reached, in the quest for better reliability, where a balance is struck between the increasing costs of development and production and the savings made in maintenance personnel and spares. The position of this optimum balance point depends upon a number of factors. The cost of improving the reliability of resistors, for example, may soon be recovered since they are manufactured and sold by the million; for travelling-wave tubes, however, this is not so.

RELIABILITY AND MAINTENANCE

Within reasonable limits increased reliability pays off in reduced costs of maintenance personnel, spares and test equipment. More reliable components, however, tend to generate more complex equipment, using more components and thus to create an overall reliability problem again. Increased reliability of components has enabled automatic test equipment to be developed. This is discussed further in Chapter 10.

In the Royal Air Force, studies show that at station and squadron level total maintenance costs for aircraft are about 40% of the total station costs, and that one-quarter of this is in respect of electronic equipment. These figures include personnel and spares. If the reliability of all aircraft electronic equipment were doubled, it is estimated that the consequent saving in personnel and spares costs would be just under 2% of the total cost of running the Royal Air Force. This estimate ignores the increased cost of providing the electronic equipment which is twice as reliable, and is therefore highly unrealistic, but it does give an idea of the magnitude of the sums involved and of the possible savings. Even modest improvements in reliability can produce savings of the order of millions of pounds.

HOW RELIABILITY IS MEASURED

So far we have discussed some aspects of reliability but have avoided the question 'What is meant by reliability?' Reliability is connected with failure and there is a straightforward relationship between failure and reliability. Let us consider failure first.

FAILURE RATE

Failure of a component, unit or equipment occurs when it fails to perform its intended function under operating conditions. A failure may be loss of output, or a change in output to a value outside the specification limits. Specification limits are necessary because, in a complex equipment, the tolerances in the various circuit outputs must be small so that errors accumulated in a large number of circuits do not become too great.

The **failure rate** of a component or equipment is the number of failures divided by the time during which they occur. Thus it may be found that in a piece of equipment 33 resistors fail per year. This figure is not helpful as it stands since the total number of resistors is not mentioned, and failure rate is often quoted as percentage of failures in a given time. Suppose there are 1000 resistors; the failure rate can then be stated as 3.3% per year. If in the same equipment there are 120 switches, and every year six of these fail, the failure rate for switches is 5% per year. This method of expressing failure rates enables the reliability of resistors and switches to be compared directly as far as their use in this particular equipment is concerned. Note, however, that in a different equipment these figures would not apply; the switches may have to carry more current, for example, and thus would probably show a higher failure rate.

The failure rate for components is often expressed as percentage failures per 1000 hours, but for relays and switches it may be expressed as percentage failure per 1000 operations.

Study of electronic components and equipments shows that under normal conditions the failure rate varies as in Figure 2.1.

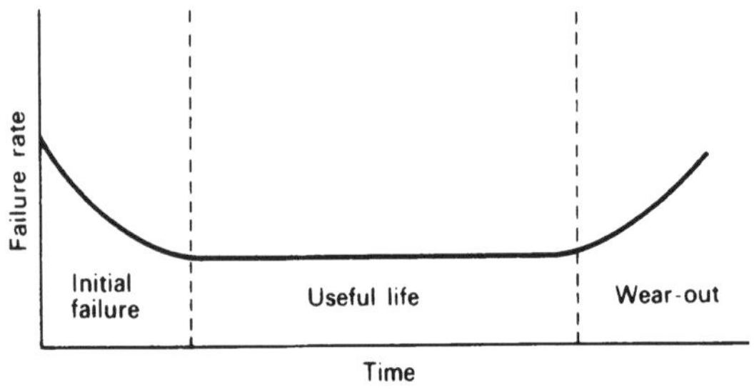

Figure 2.1 Variation of failure rate with time

There is an initial period of high failure rate during what is known as the **burn-in, initial-failure** or **infant-mortality** period, during which time the failure rate falls to a steady value. The initial-failure period is the fairly short period during which failure due to faulty manufacture occurs. When these failures are finished a period of low, constant failure rate follows.

The constant failure-rate period is known as the **useful life**, and during this period failures are random in nature. This means that failures are due to chance alone, and that the chance of a failure occurring is the same at all times during the useful life. This is the period of most interest to us, for it covers the normal life-span of the equipment or component.

Finally, the **wear-out period** shows a rising failure rate as the effects of age, usage and chance combine to give an increasingly unreliable equipment.

These three periods apply to motor cars as well as to electronic equipment, and help to determine such factors as the best time to sell or buy, and secondhand values.

RELIABILITY MEASURED AS A PROBABILITY

When failures are due to chance alone, and thus occur during the period known as the useful life of the component or equipment, the reliability is also dependent on chance; it is dependent on the chance of failure occurring. The term **reliability** expresses the chance that a component or equipment will function normally for a specified period of time; it is the probability that it will perform its proper function under normal operating conditions for a specified period of time.

Probability expresses the chance of a certain event happening. It may be, for example, that over a large number of years it is found that it snows on 1st January one year in eight. The chance, or probability, that it will snow on 1st January next year can then be said to be one in eight, or $\frac{1}{8}$, or 12½%. Probability can be quoted as a number or as a percentage. It can never be greater than 1, or 100%, because these figures represent absolute certainty that an event will occur.

If we say that an equipment has a reliability of 70% for a 100 hour period, we mean that if we operate it for 100 hours, seven times out of ten it will function correctly with no failures.

Reliability as defined above involves **time.** A statement that a certain equipment is '95% reliable' is meaningless unless the time is quoted.

There is, as has been mentioned, a close relationship between reliability and failure rate, and it is a very simple relationship. Taking our equipment which is 70% reliable over a 100 hour period, this means that seven times out of ten it will function properly when operated for 100 hours.

The remaining three times out of ten it will fail. The failure rate of the equipment is thus 30% per 100 hours. If the reliability had been expressed as 7/10 or 0.7 per 100 hours, the failure rate would be expressed as 3/10 or 0.3 per 100 hours. It can be seen that:

$$\text{Reliability} = (1 - \text{Failure Rate})$$

When reliability and failure rate are measured as percentages, the expression becomes

$$\text{Reliability} = (100 - \text{Failure Rate})$$

RELIABILITY MEASURED AS MEAN TIME BETWEEN FAILURES (M.T.B.F.)

The reliability figure mentioned so far is not the ideal for practical use, since the probability that a piece of equipment will perform successfully depends upon the conditions under which it is operating, the performance expected of it and the time of operation. More useful to the user is the average time that an equipment will run between failures. This time is known as the **Mean Time Between Failures (m.t.b.f.)**. The m.t.b.f. is also closely related to the failure rate as may be seen by the following example. Suppose a unit consists of 1000 resistors whose failure rate is 3.3% per year. Then on average 33 resistors will fail per year, and, again on average, a time of 1/33 of a year will elapse between resistor failures, i.e. about 11 days. So the unit may be expected to run about 11 days before a resistor fails, and this is its m.t.b.f.

It must be remembered that reliability, failure rate and m.t.b.f. all depend on chance – on the chance of a component unit or equipment failure. It would be a more accurate concept to say that if the above unit is operated for a very long time, on average it would fail every 11 days. In a short period the failures may be very haphazard, and the times between failures may run 6, 18, 14, 7, 10, 12, 9 days, only averaging 11 days over a long period.

If there are n components of a particular type, e.g. resistors, capacitors or transistors, and the average failure rate is f per hour, then nf components will be expected to fail every hour. The m.t.b.f. (m) is then:

$$m = \frac{1}{nf} \text{ hours}$$

Notice that the units in this formula must be consistent; if the m.t.b.f. is required in hours, the failure rate must be expressed in number of failures per hour, **not** per 1000 hours.

THE RELATION BETWEEN RELIABILITY AND M.T.B.F.

During the useful life of an equipment or component the failures occur at random and the failure rate stays constant. The reliability is the probability that the unit will not fail during a specified time. Events that are governed by probability are often connected mathematically by an exponential formula. In this case, if the reliability, or the probability of success for an operating time t hours is R, and the failure rate is λ failures per hour, the formula connecting these quantities is:

$$R = e^{-\lambda t}$$

Reliability, being a probability, can be expressed as a number or a percentage.

Consider, for example, an equipment with a useful life of 400 hours and a failure rate of 3 faults per 1000 hours during this period. Then for any 24-hour period during the useful life, the failure rate is constant and the reliability is constant, the reliability being:

$$R = e^{-\lambda t} = e^{-0.003 \times 24}$$

$$= e^{-0.072} = 0.93$$

The equipment thus has a 93% chance of running for any 24-hour period without failure.

The number $e^{-0.072}$ may be calculated using exponential tables, logarithmic tables, a slide rule or a combination of these. The exponential tables given with most simple sets of mathematical tables are too simplified to be of much use; they usually start at $e^{-0.1}$ and have wide gaps between entries which require interpolation.

The number can always be calculated by logarithms in the following manner:

$$\log (e^{0.072}) = 0.072 \times \log e$$

$$= 0.072 \times 0.4343$$

$$= 0.03127$$

No.	Log
0.072	$\bar{2}.8573$
0.4343	$\bar{1}.6378$
0.03127	$\bar{2}.4951$

$$\log (e^{-0.072}) = -0.03127$$

$$= \bar{1}.96873$$

whence $e^{-0.072} = 0.9305$

The reliability of the equipment for the whole period of its useful life of 400 hours is:

$$R = e^{-0.003 \times 400}$$

$$= e^{-1.2} = 0.30$$

It has therefore a 30% chance of running for 400 hours without failure.

If we deal with similar components instead of equipments, the failure rate λ used above is the number of component failures per hour, and is the failure rate per component multiplied by the number of components. It is, in fact, the quantity **nf** that we used in the last section, where we showed that m.t.b.f.:

$$m = \frac{1}{nf}$$

So, if λ is the number of failures per hour, the m.t.b.f. must be:

$$m = \frac{1}{\lambda} \text{ hours}$$

It is necessary when using this formula to express the units correctly. If for example we have 4000 components with a failure rate of 0.02% per 1000 hours, the average number of failures **per hour** is:

$$\frac{0.02}{100} \times \frac{1}{1000} \times 4000$$

$$= 8 \times 10^{-4} \text{ failures per hour}$$

The m.t.b.f. is therefore:

$$m = \frac{1}{8 \times 10^{-4}} = 1250 \text{ hours}$$

As another example, consider a unit consisting of 800 transistors, each with a failure rate of 0.05% per 1000 hours, and with a useful life of 2000 hours. We wish to know its reliability over any 10-hour period, and over its useful life.

First express the failure rate in number of failures per hour. This is:

$$\lambda = \frac{0.05 \times 800}{100 \times 1000}$$

$$= 0.0004 \text{ failures per hour}$$

The reliability for a 10-hour period is therefore:

$$R = e^{-\lambda t} = e^{-0.0004 \times 10}$$

$$= e^{-0.004} = 0.996 \text{ or } 99.6\%$$

The reliability for the useful life of 2000 hours is:

$$R = e^{-0.0004 \times 2000}$$

$$= e^{-0.8} = 0.45 \text{ or } 45\%$$

In dealing with an equipment or system, λ is more appropriate than nf, being simply the average number of faults per hour. Hence the reliability for a time t is:

$$R = e^{-\lambda t}$$

$$= e^{\frac{-t}{m}}$$

where m is the m.t.b.f.

A graph of reliability against time is shown in Figure 2.2.

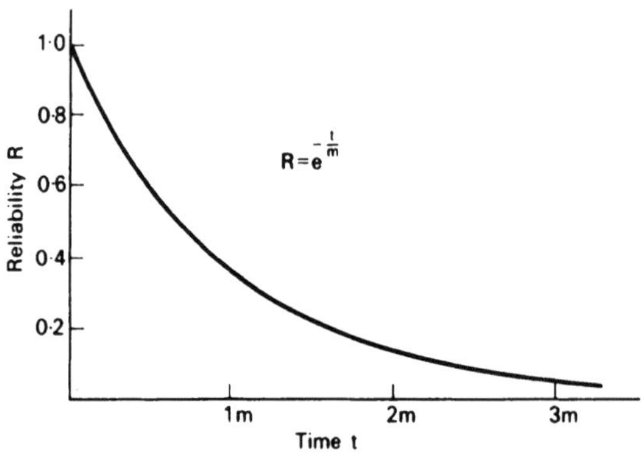

Figure 2.2 Reliability curve

The time, t, can be measured from an arbitrary point within the useful life period of the equipment, but must not extend into the wear-out period where the formula does not hold.

This exponential curve will be recognized as being similar in shape to the decay curve of current in a circuit containing a capacitor discharging through a resistor, and the m.t.b.f. is analogous to the time constant in a CR circuit. As time increases the reliability decreases and, when t = m, the m.t.b.f., the reliability is only 36.8%. Thus an equipment with a mean time between failures of say 100 hours has only a 36.8% chance of running 100 hours without failure.

As an illustration of the use of the formula, imagine a computer with a m.t.b.f. of 50 hours. The probability of a successful 10-hour run is then:

$$R = e^{\frac{-t}{m}} = e^{\frac{-10}{50}}$$

$$= 0.819 \text{ or } 81.9\%$$

The probability of a successful 2-hour run is:

$$R = e^{\frac{-2}{50}}$$
$$= 0.961 \text{ or } 96.1\%$$

In practice we are usually more interested in the small portion at the upper end of the reliability curve, and over this portion the graph may be taken to be a straight line as shown in Figure 2.3.

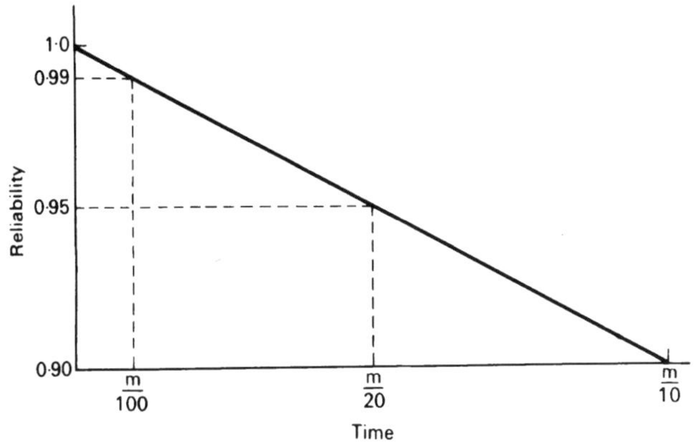

Figure 2.3 Upper part of reliability curve

If it is assumed that the graph is a straight line over this small section, reliability is proportional to time. For a time equal to one-tenth of the m.t.b.f., the reliability is 90%, and for lesser times simple proportion may be used to calculate the reliabilities without performing calculations involving exponentials. Examples are shown on the graph of Figure 2.3.

AVAILABILITY

As far as the user of an equipment is concerned, the m.t.b.f. is not the only measure of importance. When an equipment is attended, repair of a failure can be started as soon as it occurs; in other cases, such as in an aircraft in flight, repair cannot usually be undertaken immediately.

In the latter case the m.t.b.f. is of great importance, but in both cases the user is interested in how long the equipment will remain unserviceable while repairs are being carried out. The term which is used in this connection is the **availability** of the equipment, and this is the proportion of its operating time that the equipment is in working order. It is also called **maintainability** in American literature, or **uptime ratio.** The availability thus depends on the **mean time to repair** a fault **(m.t.t.r.)** as well as on the mean time between faults, and it is given by:

$$\text{Availability} = \frac{\text{m.t.b.f.}}{\text{m.t.b.f.} + \text{m.t.t.r.}}$$

If the m.t.t.r. can be reduced, availability will increase and the utilization of the equipment will be more economical. The m.t.t.r. can be reduced by a good design, and good maintenance policy, which enable the time taken to carry out routine procedures and to diagnose and repair a fault to be kept low. Good design will incorporate such features as providing suitable monitoring and test points, making components accessible, and using plug-in chassis or printed circuits. The m.t.t.r. can also be reduced if the method of fault location is speeded up, and this book is mainly concerned with methods by which this can be done.

The m.t.b.f. of an equipment may be estimated from a knowledge of the number and types of its components, in a manner which is described later, but it is much more difficult to calculate the m.t.t.r. since this depends on a number of factors which are difficult to measure. Some of these factors are discussed in Chapter 4.

THE RELIABILITY OF ELECTRONIC COMPONENTS

As has been mentioned, the failure rate of components depends upon the type of equipment in which they are used, i.e. it depends on the environment, as well as upon the inherent proportions of the components themselves. Components in a computer, static in the fairly constant atmosphere and conditions of a computer room, will have a different failure rate from the same components used in a missile. For this reason one finds widely different failure rates quoted for components. Table 2.1 gives figures applicable to a certain computer.

Table 2.1: Component failure rates
(Per cent per 1000 hours)

Valves	0.228
Transistors	0.008
Semiconductor diodes	0.004
Resistors	0.0004
Capacitors	0.0016
Transformers	0.009
Relays	0.004
Plugs and Sockets	0.0008
Lamps	1.41

The failure rate of all components usually increases with increase of temperature, mechanical stress and electrical stress; an environmental factor

MB

can sometimes be used to allow for this. For example, the failure rates of the ground-based computer components quoted in Table 2.1 should be multiplied by 5 to 10 times if the same components are used in an aircraft installation.

THE RELIABILITY OF A SYSTEM

To predict the reliability of a system the failure rates of each of the components must be known, or estimated, and the overall failure rate deduced as shown in Table 2.2. The choice of failure rate for each component is not easy, as has been mentioned, but a figure derived from information concerning failure of components in a similar system working under similar conditions is likely to be most useful. Table 2.2 might therefore represent the calculations for a computer using the same type of components in the same sort of conditions as apply to Table 2.1.

Table 2.2: Data for reliability calculation

Component	Number	Failure rate % per 1000 hours	Total failures per 1000 hours
Transistors	6000	0.008	0.480
Semiconductor diodes	25000	0.004	1.000
Resistors	30000	0.0004	0.120
Capacitors	6000	0.0016	0.096
Transformers	150	0.009	0.014
Relays	100	0.004	0.004
Plugs and Sockets	2000	0.0008	0.016
		TOTAL	1.730

The total number of failures per 1000 hours is on average 1.73, and the m.t.b.f. for the computer is therefore:

$$\text{m.t.b.f.} = \frac{1000}{1.73} = 578 \text{ hours}$$

MAJOR CAUSES OF UNRELIABILITY

As may be imagined, failure of an electronic equipment may be due to a number of causes in addition to the failure of components. Negligence, faulty design, omissions, unsatisfactory parts and poor workmanship all contribute towards failures. The list below shows some of the causes of failure, in order of importance:

1. Inadequate engineering design.
2. Component failures.
3. Poor handling in operation or maintenance.
4. Poor workmanship and lack of checking.

Inadequate engineering design covers such factors as wrong choice of component, equipment not being able to stand up to its environment, use of unsuitable materials and complexity of equipment.

Among **component failures**, valves are the most unreliable by a long way, and their replacement by transistors, where possible, leads to much increased reliability. In the Royal Air Force, capacitors follow valves as being the next most unreliable components, followed, in order, by variable resistors, relays, plugs and sockets transformers, fixed resistors, switches, and inductors.

Poor handling in operation or maintenance includes such faults as using the equipment in abnormal or poor conditions, and faulty maintenance, some of which may be due to lack of a proper guide to maintenance or even the lack of a proper diagram.

The fourth item, **poor workmanship and lack of checking**, is self-explanatory, and it covers all the reasons why faulty equipment is passed into service instead of being rejected during production checks.

FAULT INFORMATION

In order that information concerning failure rates and causes of failure is available to the equipment designer, it is essential that information concerning faults is fed back from the users of the equipment. Only by having this knowledge available can the reliability of newer systems be estimated, even very roughly, and only by this feedback of information can re-design of the equipment be undertaken in order to cure faults which are not random. The technician is an important link in this feedback of information. It is he alone who knows exactly what routine procedures have been carried out and the exact nature of faults and the associated repairs. Unfortunately, the reporting procedure often involves what appears to be tedious and apparently wasteful paperwork to the technician, who may be remote from the person who uses the information. A good maintenance organization would provide a means of showing the technician how this recorded information is used, thus demonstrating its importance to him and encouraging him to be clear and accurate in his reports.

Testing of components in laboratory conditions gives a guide to their reliability, but it is only a guide. The reliability varies with the environment and laboratory conditions do not equate to operational conditions. Motorcar manufacturers find that some faults in new models, which their design and testing staff never experience, appear as soon as the cars are on sale to the public, and some manufacturers release new models to certain members of the public before they are generally put on sale, to see if unsuspected faults develop.

HOW RELIABILITY MAY BE IMPROVED

Most of the methods for reducing the effects of environment, which are described in Chapter 3, will also improve the reliability of equipment and components under normal conditions. In addition to these methods there are some other ways in which reliability may be improved, the most important being attention to circuit design, incorporating redundancy, derating components, constructional developments such as integrated circuits, and cooling.

CIRCUIT DESIGN

There is a very great range of possible improvements which can be brought about by attention to detail when circuits are designed. Maximum efficiency is most desirable, because the wasted power in an inefficient circuit appears in the form of heat which must be dissipated. Low heat dissipation saves weight, size and complexity, and reduces the high temperature effects on components. Low heat dissipation is one of the reasons why transistors have replaced valves in modern circuits, except for a few special applications.

Circuit and general equipment design must also build in the experience gained by feedback of fault information from similar equipment used operationally.

REDUNDANCY

When all possible means of improving reliability have been employed, following the lines given in this chapter, it is still possible further to improve the reliability of an equipment by the process of **redundancy**. This method is particularly applicable to systems which cannot be repaired e.g. airborne equipment in flight or satellite equipment. Redundancy consists of providing alternative paths for some or all of the functions of the equipment, and is achieved by duplicating or triplicating certain parts. Redundancy may be at component, unit or equipment level. At component level redundant components are usually permanently in circuit, and the circuit design is such that failure of all but the last serviceable component will not make the equipment inoperative. Above component level redundancy usually involves some sampling mechanism, whereby a selector calls on another unit or equipment if the first one fails. This means that more components not directly related to the function of the equipment have to be included, and their reliability will contribute to the equipment's overall reliability. An example of equipment redundancy is the automatic landing system fitted to some modern airliners, which is triplicated to achieve an exceedingly high standard of reliability.

At component level there are two general methods of achieving redundancy – series and parallel. Series redundancy (Figure 2.4) involves connecting two components in series, each one alone being sufficient for the circuit. If one component fails to a short circuit, the other will continue to let the circuit operate. It is clear that this method is only suitable where the most probable failure is to a short circuit condition.

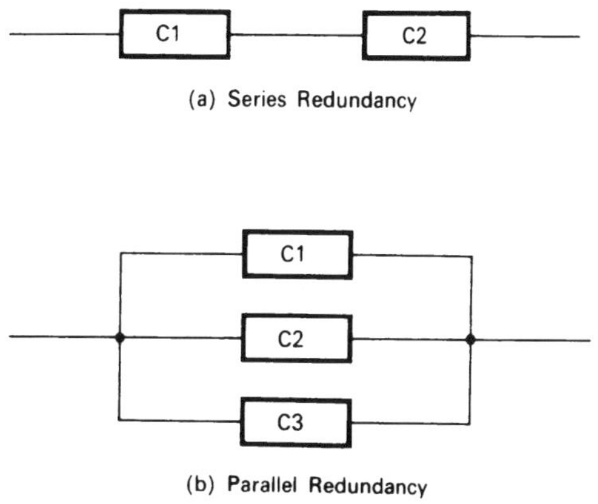

Figure 2.4 Redundancy

Parallel redundancy is also possible, (Figure 2.4) particularly where components have a high probability of failing to an open-circuit condition. It does have hidden dangers, for if one parallel path fails, the consequent change in circuit conditions may degrade the general performance and may overload other components. Parallel redundancy is more useful at unit or equipment level, particularly when a switch brings in a parallel unit in the event of failure; failure at the switch itself has then also to be considered.

As an example of the increase in reliability that can be achieved by parallel redundancy, imagine an equipment with an m.t.b.f. of 500 hours. For an operating time of 80 hours its reliability is therefore:

$$R = e^{\frac{-t}{m}}$$

$$= e^{\frac{-80}{500}} = e^{-0.16}$$

$$= 0.852 \text{ or } 85.2\%$$

If two similar equipments are permanently connected in parallel, the reliability of the combination is the probability that the combined system will not fail for 80 hours; this must be calculated from the failure rate. The probability that one equipment will fail in 80 hours is 1 – 0.852 = 0.148. The probability that both equipments will fail in this time is 0.148 x 0.148 = 0.0218, and the combined reliability is thus:

$$R = 1 - 0.0218$$

$$= 0.978 \text{ or } 97.8\%$$

If three similar equipments are connected in parallel, the probability that all three will fail in 80 hours operating time is 0.0218 x 0.148 = 0.0032, and the combined reliability is:

$$R = 1 - 0.0032$$

$$= 0.997 \text{ or } 99.7\%$$

DERATING

Another method of improving the reliability of components is by operating them below their normal power, so that heat dissipation is reduced and hot-spots are avoided. Valves, resistors, and capacitors show significant improvement when they are derated.

A very rough rule for resistors is that the failure rate is halved if their dissipation is halved. A resistor with a failure rate of 0.01% per 1000 hours under normal operating conditions would have a failure rate of approximately 0.005% per 1000 hours when run at half its rated dissipation. The gain in reliability for capacitors is even more marked; a mica capacitor with a normal failure rate of 0.03% per 1000 hours may reduce to 0.004% per 1000 hours when run at half its rated voltage.

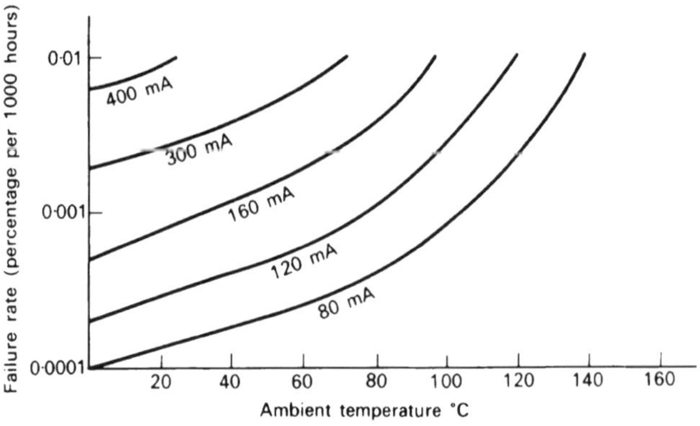

Figure 2.5 Reliability of silicon diodes through derating

Figure 2.5 is a graph showing the improvement in reliability which is obtained by derating a certain type of silicon diode. Not only does the graph show how the failure rate falls as the current through the diode is reduced, but it also illustrates the rise in failure rate as the surrounding temperature increases.

SOME CONSTRUCTIONAL METHODS WHICH IMPROVE RELIABILITY

Many methods have been developed which are intended to protect electronic equipment and components against the effects of environment, and hence to improve reliability. The worst effects are those due to humidity and vibration, and the four main recent developments in construction have been encapsulation, printed circuits, wrapped joints and integrated circuits. There have also been tremendous developments in components themselves which make them less sensitive to environment. The development of semiconductor diodes and transistors has reduced the problem of mounting in order to minimise the effects of shock and vibration, because they are small; also, because they dissipate little heat neighbouring components do not suffer from high temperatures, although the dissipation of what heat there is from within the transistor creates another problem. High-stability resistors and ceramic capacitors are less affected by working conditions than older types of component.

Encapsulation or potting of circuits and components provides a great deal of protection against humidity, shock and vibration. Encapsulation is done by covering the component with a liquid plastic resin that sets hard on cooling. It is essential, however, that the potting is done carefully, otherwise a partial seal may let in water which remains trapped and may cause more damage than if potting had never been carried out.

Printed circuits, embracing printed wiring and printed components such as resistors, inductors and capacitors, permit no wiring errors and eliminate many connections which can contribute towards failure. They stand up to vibration well.

Wrapped or solderless joints are formed by winding, under controlled tension, seven or eight turns of soft wire round a hard metal peg with sharp corners. The corners bite into the wire, and make a good electrical and mechanical connection. Specialized but relatively simple tools are needed to make and strip wrapped joints.

The reliability of wrapped joints is exceedingly high. In a certain data-processing equipment containing nearly 900,000 soldered joints there were 5 failures in four years, representing a failure rate of 0.00003% per 1000 hours. In an airport message-switching system containing 6.5 million wrapped joints there was only one failure in over two years. This represents a failure rate of less than 0.000001% per 1000 hours, showing that wrapped joints give the lowest figures of failure of all known techniques of connection.

INTEGRATED CIRCUITS

With the invention of transistors it became possible, in theory at least, to reduce the size of electronic devices to very small dimensions indeed, because vacuum-sealed envelopes, and spacing between electrodes to provide insulation,are not necessary with semiconductor devices. The practical difficulty is in handling microscopic pieces of material, and development has allowed many interconnected circuits to be constructed from one small piece of semiconductor material with no external connecting wires whatsoever. Circuits of this type are known as **integrated circuits**, and they are manufactured in two forms: thin-film circuits and monolithic circuits.

Thin-film circuits are made by depositing, by evaporation or sputtering, metals and dielectrics on a glass sliver, where they form elements such as resistors, capacitors and conductors. The circuit shape is achieved by using a thin metal mask, or by photographing the pattern on the surface and etching away the non-exposed parts. Films are typically one-millionth of an inch thick, and components can be packed at a density of 100 to 1000 per cubic inch. Diodes and transistors for these thin films are manufactured with all electrodes on one face of the basic semiconductor material, inverted on to the thin film, and welded into place.

Monolithic circuits are made from a thin, flat piece of silicon, which is subjected to a series of masking, etching and diffusion processes. These processes form components, such as resistors and transistors, in a thin surface layer of the material, connected by other pieces which have been formed into conductors or insulators, as appropriate, by diffusion and etching. Again the circuit configuration is produced by photographing, on a greatly reduced scale, a pattern drawn with pen and ink. A single slice of silicon one square inch in area can contain over 500 complete circuits which are all etched and diffused in one continuous process, and the packing density of components where monolithic circuits are used can reach 1000 to 10,000 per cubic inch.

The elimination of separate joints, connecting wires, seals and insulators which is achieved with integrated circuits produces a very high order of reliability, although faults can still occur in the numerous circuit connections which have eventually to be made. The recent advance in reliability that has been achieved is shown in Figure 2.6, and the trend in the near future is that even greater reliability can be expected.

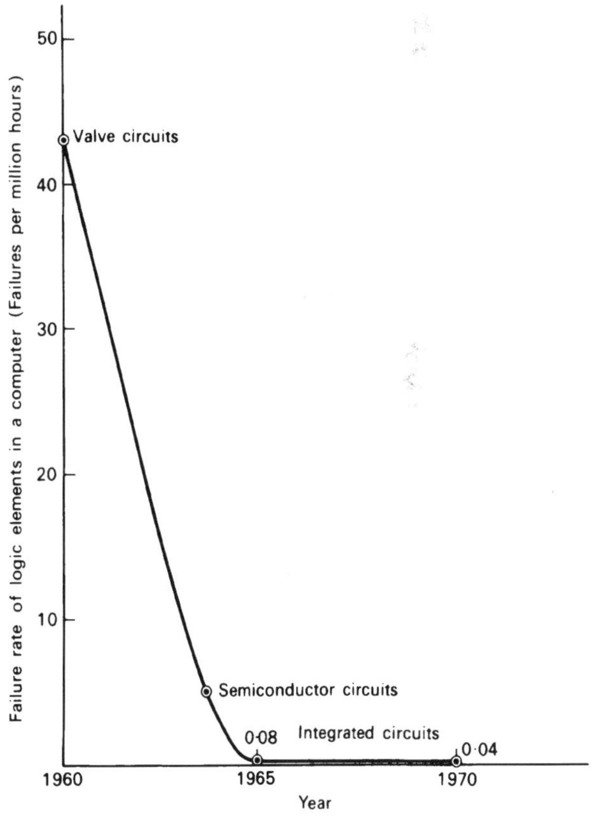

Figure 2.6 Improvement in reliability with development of integrated circuits

However, this trend towards greater reliability of components should not be taken to mean that maintenance and maintenance technicians will no longer be required in the future. As has been mentioned already, greater reliability of components leads to more complex equipment so that the overall problem of maintenance remains, even though its form may change in detail. This new development in the field of microelectronics will undoubtedly produce its own special maintenance problems.

COOLING

Cooling of all heat-producing components is essential to good reliability. There are many factors to be taken into account in designing the best method of cooling components, and allowing for the amounts of heat lost by conduction, convection and radiation. Components themselves pose different problems. A valve for example has a large surface area which can be used for radiation and convection cooling, whereas a transformer generates a lot of its heat inside a mass of metal and must rely largely on conduction for heat dissipation. Encapsulation makes the problem of cooling very difficult because the plastic resin used is a poor conductor of heat; thus components with large heat dissipation are not suitable for potting. The replacement of the valve by the transistor is solving a lot of cooling problems, since valves are the components which produce most heat in electronic equipment. However, modern equipment is expected to operate in more difficult environments than has been the case in the past e.g. satellites on re-entry reach very high temperatures and equipment on the moon is subject to very low temperatures. The problems of cooling, and heating, still remain.

SUMMARY

The **failure rate** of a component or equipment is the number of failures divided by the time during which they occur, and it is often expressed as percentage failures per 1000 hours. The failure rate normally varies during the life of the component or equipment from a high initial failure rate, through a constant rate during the useful life, to a terminal high-failure rate wear-out period.

Reliability is the probability that no failure will occur for a specified period of time, and Reliability = 1 − failure rate .

The **mean time between failures** (m.t.b.f.) is the average time that an equipment will run between failures. The reliability, R, for a time t hours is

$$R = e^{-\lambda t} = e^{\frac{-t}{m}}$$

where m is the m.t.b.f. and λ is the average number of faults per hour.

The **availability** of an equipment is the proportion of its operating time that it is in working order.

$$\text{Availability} = \frac{\text{m.t.b.f.}}{\text{m.t.b.f.} + \text{m.t.t.r.}}$$

where m.t.t.r. is the mean time taken to repair a fault, including the time awaiting spares.

The reliability of a system may be predicted by summing the failure rates of its components.

The major causes of unreliability, in order of importance, are: inadequate design, component failures, poor handling, poor workmanship.

Reliability may be improved by: good circuit design, incorporating redundancy, i.e. by providing alternative paths for some or all of the equipment's functions, derating, i.e. operating components below their normal power, using improved constructional methods such as encapsulation, printed circuits, wrapped joints and integrated circuits, and paying attention to cooling.

CHAPTER 3 COMPONENTS AND RELIABILITY

INTRODUCTION

Many of the causes, such as design faults, contributing to failure of an equipment cannot be foreseen, or calculated in advance, and thus cannot be used to predict the reliability of a system. It is for this reason that failure rates nearly always relate to the useful life period of components, for, even though the failure rate in this period changes with the environment of the equipment, at least some attempt at prediction can be made. Information concerning the failure rate of a new equipment containing new components can be obtained by two main methods. These are feeding back information concerning faults after the equipment has entered service, and testing components under artificial conditions before the equipment is manufactured.

Both of these methods have their uses, and they both give information which can be used to improve the design or manufacture of the equipment. Laboratory testing of components enables the effects of a wide range of environment to be studied - temperature, humidity, pressure, and so on. Relying solely on the feedback of fault information is not possible with completely new equipment, and some prediction must be made if only to ensure that faults will not be so frequent that the equipment never works. Feedback of fault information is therefore a method of modifying a much cruder forecast of reliability.

Laboratory testing of components does, however, have its problems. Many modern components are so reliable that thousands of components have to be tested for thousands of hours before even an approximate failure rate can be deduced. For example, the failure rate of resistors in a certain equipment may be 0.0004% per 1,000 hours. To produce only one fault in a year's testing would require 28,300 resistors. The scale of testing is limited and this means that not as much information, covering different conditions, can be made available as we would like. Occasionally laboratory tests can be devized which speed up the failure rate, for example by running transistors at a higher temperature, to give more failures in a given time. Interpreting the results in terms of components at ordinary temperatures is again, however, uncertain.

This chapter describes very briefly how components are tested for the commonly met conditions of humidity, temperature, vibration, shock, acceleration, pressure and radiation, and what the common faults in components are found to be as a result of these tests. The effects of humidity, temperature, shock, vibration, pressure and radiation on common components are also described.

HOW TESTING IS CARRIED OUT

It is not possible in this book fully to cover the enormous range of tests which are carried out on components under varying conditions, and only certain aspects will be mentioned. As an illustration, capacitor dielectrics may be tested for permittivity, loss angle, surface and volume resistivity, dielectric strength and insulation resistance. Clearly not all of these are of interest to the maintenance technician.

Components may be tested under varying conditions of humidity, temperature, vibration, shock, acceleration, pressure and radiation.

The reaction of components to varying humidity is tested in a box which is temperature-controlled by a thermostat, and whose humidity can be altered by injecting water vapour from a boiler. A fan gives an even distribution of temperature and humidity within the box, and the variation can be made cyclic if desired. Humidity testing is necessary when equipment is to be operated in different climatic conditions, from polar to tropical, and when it must retain its specification within acceptable limits.

Variation of performance with temperature is tested in a similar lagged box, or perhaps in the same one. A refrigerator can supply the box with cold air, a heater with hot air; for higher temperatures the box itself may be in an oven. Temperature testing is necessary to simulate the effects of different climates or environments, such as high-flying aircraft.

Vibration testing is carried out in a machine that may be electrically or mechanically driven. The electrical machine is a robust form of loudspeaker in which the normal sound-producing cone is replaced by a spigot which drives a table to which the test piece is attached. It is more versatile than a mechanical vibrator because, by varying the oscillator output, the frequency and amplitude of vibration may be varied at will. The oscillator also drives a stroboscope which shows any effects in slow motion. If a mechanical resonance is found within undesirable frequency limits, the cause can be found and the defect cured by stiffening and weighting. The effects of vibration are important because it is present in all forms of transportation, and also on working surfaces near heavy machinery.

The effects of shock are investigated by subjecting the equipment or components to a cam-operated bumping machine driven at 100 to 200 rev/min. This machine gives impact shocks of approximately 20g retardation or acceleration and components are subjected to about 3,000 shocks. Shock testing supplies information on the effects of bad handling, such as at docks, and on operation in military vehicles, warships and aircraft.

Acceleration tests are carried out by means of a small centrifuge; alteration of the length of the rotating arm and speed of rotation produces different values of acceleration. These tests are used where components are liable to be subjected to high values of acceleration in operational use, e.g. in high-speed aircraft or missiles. Acceleration in these circumstances may for example switch on relays when they are not required.

The effects of low pressure on insulation breakdown, tracking at high voltages and the efficiency of sealing are investigated by using bell-jars with rotary vacuum pumps.

Testing the effects of nuclear radiation is not a common procedure, but components have been subjected to radiation from nuclear power plants and some general results obtained.

COMMON FAULTS IN COMPONENTS

Vibration testing has shown up many weaknesses in mechanical design of units, and it is important that mechanical resonances are detected and made to occur outside the sensitive frequency range. With airborne electronic equipment lightness is essential, and modern practice is to start with a very light, but weak, structure, and then strengthen it after vibration testing. Mechanical resonances occur with all mounting methods and some faults occur because the methods of component mounting result in broken leads. Sealed equipment is found to be more reliable than unsealed.

Thermionic valves have the highest failure rate of all components, and the commonest valve faults are due to excessive temperature, mechanical shock or vibration, and loss of cathode emission. Valves designed to operate at a certain maximum temperature may exceed this temperature because they are located too near other hot components, or because the cooling arrangements are not satisfactory due to lack of circulation, shielding and so on.

Fixed resistors are comparatively fault-free, and the faults that do occur are mainly a change in resistance value due to heating of the material, and open circuits due to faulty connections. Cracked-carbon resistors are more sensitive to climatic conditions than the carbon-composition type. Wirewound resistors suffer most from open circuiting due to faulty connections with the fine wire that is used, or corrosion of the wire. Variable carbon-composition resistors are prone to noisy contacts produced by dust and grit on the wiper track; tracking can also occur in humid conditions.

Fixed capacitors suffer most from breakdown of insulation caused by moisture penetrating the sealing, and silver-mica capacitor failures are often due to the deposition of a metallic film across the dielectric. Faults in the electrolyte are the major causes of failure in electrolytic capacitors; drying-out at high temperatures, freezing at low temperatures, and corrosion at all temperatures. Electrolytic capacitors are also prone to failure because of the possibility of connecting them in circuit the wrong way round; it is not always obvious which is the correct connection. Particular care must therefore be taken to check connections. Another common source of failure is breakdown due to excessive voltage between the plates. This is most common in small equipment, where the electrolytic capacitor specified is often made as small as possible, and its tolerance is only just within the required operating voltage. Variable capacitors are very reliable components and faults are rare. The main troubles arise with air-spaced variable capacitors which, if they are not protected, are liable to receive mechanical damage which changes or closes the small gap between the plates. A deposit of dust or moisture between the plates causes similar faults.

Faults in r.f. cables are rare, but those that occur are mainly caused by moisture, corrosion, overloading with power or voltage, mechanical damage causing a change in r.f. properties, and terminations. Connection to plugs and sockets often causes difficulty due to the combination of a fine centre wire, an insulator of low melting-point and heavy braiding. Care in handling r.f. cables can often eliminate difficult fault location problems at a later stage. Ordinary connecting wire is fairly fault-free, and loss of insulation around the wire is the main trouble, usually caused by moisture.

The commonest fault in plugs and sockets is breakage of soldered joints due to pulling on the leads. Contact troubles are fairly rare if covers are used when the plug is not connected for long periods; gold-plated contacts also reduce contact faults. Plugs and sockets used at r.f. are generally not as well-designed electrically or mechanically as l.f. and d.c. types, and mechanical failures are fairly frequent.

Relays suffer from corrosion of fine wires and consequent open-circuiting, and from metal fatigue in the armature spring. Dirty contacts give trouble and must be cleaned frequently; this fault is often due to the action of chemicals produced by sparking in an enclosed space. Switches, particularly wafer switches, are liable to fail because of poor contact, and again frequent cleaning is necessary.

Modern fully-sealed transformers are generally fault-free unless they are overloaded, perhaps through failure of another component, but impregnated transformers have more faults. Moisture is again the biggest offender because it weakens the insulation.

Meters generally have a long life when they are treated properly, but occasionally a sticky movement occurs, caused by corrosion. Chapter 10 deals with the care and use of test instruments including meters.

THE EFFECTS OF ENVIRONMENT ON COMPONENTS

Equipment that has to operate in difficult conditions such as high humidity, high or low temperatures, high or low pressure, or vibration will naturally have a higher failure rate than that which is used in stable surroundings. Not only does the complete failure of components become more frequent, but small changes in a large number of components result in the equipment as a whole becoming liable to deteriorate in performance until, without a catastrophic failure, it fails to meet its specification limits.

HUMIDITY

Humidity is the factor which affects the operation of electronic equipment most, particularly when high temperature exists at the same time, as in tropical conditions. Moisture has two main effects: the insulation resistance of tag strips, wafers and other insulators is reduced because of the formation of surface film and the absorption of water; and water may form an electrolyte between dissimilar materials thereby producing a spurious voltage by galvanic action. There are also other effects of humidity such as corrosion of plated metals and growth of fungus. Fungus contains a high percentage of water, so that if it grows on insulation, for example, it provides a high-resistance path for current.

Prevention of these effects of humidity is carried out by choosing insulating materials which do not support a surface film of water and do not absorb it. Glass, quartz and steatite porcelain easily support a surface film; polystyrene and silicones prevent a water film from forming. On the other hand, glass does not absorb water and neither does polystyrene, many ceramics or polyethylene. Cellulose materials should be avoided in conditions of high humidity. Some plastics support the growth of fungus, but ceramics, mica, glass, nylon and polyethylene, for example, are not subject to fungal attack.

Maintenance technicians should be aware of the effects of humidity, and be on the look out for any possible sources of failure due to dampness. Occasionally equipment has to be used in conditions for which it is not designed, and unexpected faults can occur for this reason. Thus if unsuitable equipment is used in tropical conditions, a watch must be kept on components and materials which could absorb moisture, support a surface film of water or allow the growth of fungus.

TEMPERATURE

The most important effects of high temperatures on components are those due to softening or melting, expansion and ageing. Many of the plastics and waxes used for protection soften or melt, and protective and lubricating grease may melt. The expansion of different metals used in assemblies can lead to strains which distort the structure, and cause cut-outs to operate prematurely. High temperature accelerates all chemical processes, and rapid ageing of components is one effect which results; as has been mentioned, testing of components is often carried out at a higher temperature than normal so that the speeded-up ageing processes can be observed in a relatively short time.

The problem of heat dissipation in electronic equipment is always present, and if the equipment has to work in a high temperature this problem is made more acute. The trend towards smaller components aggravates the situation further, because the reduced surface area of miniature components makes heat dissipation more difficult. Forced air cooling can often assist in heat dissipation, but in some aircraft and missile applications the external air is sometimes hotter than that surrounding the equipment.

Some of the effects of high temperature can be overcome by using materials which will withstand the temperature to be encountered. Thus glass and ceramics are used instead of paper and certain plastics.

Equipment may incorporate special means of temperature control, such as heaters, cooling water, or thermostats, and a failure of one of these can be as disastrous as the failure of components directly concerned with the function of the equipment. The technician must be on the look out for the effects of overheating, particularly during visual inspection of the equipment. Molten or distorted insulation, and charred paint in the colour coding of resistors are common symptoms which, however, are often the results of another fault, such as a short circuit, elsewhere in the equipment.

At low temperatures components suffer from the effects of contraction, hardening and freezing. Contraction of different metals produces strains in the same way as differential expansion, but excessive cold also makes materials brittle and more liable to crack and break. Contraction, and expansion, can also affect components such as inductors and capacitors so that readjustment of critical circuits becomes necessary. Oil and grease harden and perhaps freeze, making operation of switches, controls and variable components stiff. Either lubrication has to be avoided, or special grease must be used. Some types of electrolytic capacitor freeze, and some transistors will not work at low temperatures. Batteries, primary and secondary, lose power at low temperatures, although nickel-cadmium storage batteries stand cold conditions better than most other types.

SHOCK AND VIBRATION

The effects of shock and vibration are obvious: breakage, bending and weakening, both electrical and mechanical. The methods by which these effects can be reduced are good design of the supporting structures for the equipment, and good design of the internal structure. Vibration and shock isolators, similar in principle to the rubber bush mountings of motor-car engines, are used between the equipment and its supports. Within the equipment proper layout and positioning of components will help to reduce the effects of vibration and shock.

Some manufacturers now include packaging tests, in which the equipment and the carton in which it is to be transported are subject to drop tests, stacking and tumbling tests to ensure that the package will withstand modern transport conditions. The technician will seldom meet the problem of mechanical resonance, but will inevitably meet the effects of vibration in the form of loose mechanical and electrical connections. Most routine checks of equipment include checking the security of all components and connections which are liable to suffer from vibration, and some faults will arise from this source. The technician's task is to ensure that faults occurring because of loose connection are as few as possible, by checking carefully and accurately; no special devices such as shake-proof washers must be omitted because, for example, they are apparently unnecessary in the cosy conditions of a laboratory.

PRESSURE

Low pressure is encountered at high altitudes in airborne or missile equipment, and high pressure in mines and underwater equipment. The main effect of low pressure is the breakdown of air insulation at high frequencies and high voltages, although oozing of liquid or paste through the seal of electrolytic capacitors and similar sealed components is also experienced.

Variable components must be kept free of dirt and irregularities, as these act as points for flash-over to occur. Terminals and pins relying on air insulation, as in capacitors and plugs and sockets, must have spacing adequate to prevent breakdown at high altitudes. A typical figure is that a gap that will withstand 8000 V at sea level will only withstand 2000 V at 60,000 ft.

It is rarely possible to repair a fault caused by high or low pressure on the spot; the satellite or aircraft must land, or a submarine cable must be raised, before a failure can be rectified. It should be remembered that a fault occurring because of pressure effects may not repeat itself when the equipment is tested under normal conditions. A sparking plug in a motor car provides a common illustration of a component that may appear to work quite satisfactorily when it is tested in the open air, yet which may fail to work properly under the increased pressure of the compression stroke of the engine. In these cases a knowledge of the effects of pressure is vital to fault-finding.

RADIATION

Electronic equipment may be subjected to radiation when it is used near X-ray tubes, electrostatic generators, cyclotrons and nuclear reactors; when equipment is used in missiles, spacecraft and satellites it is irradiated with cosmic rays. The term radiation covers a wide range of waves and particles, including alpha, beta and gamma rays, X-rays, neutrons and neutrinos. Many of these either interact little with matter, or can easily be stopped by shielding, but neutrons and gamma rays penetrate deeply into matter and react with it by attacking the atomic structure. Components irradiated by gamma rays and neutrons, in particular, are therefore liable to change in chemical composition.

Technicians who deal with the maintenance of equipment which is subject to radiation should be aware of the effects of radiation on materials and components so that possible failure may be anticipated and avoided, and unsuitable materials may be rejected.

Apart from the effects on chemical structure mentioned above in connection with neutrons and gamma rays, nearly all radiation can ionize materials. Insulators are therefore strongly affected by radiation, whereas conductors, which already possess large numbers of free electrons, are hardly affected at all. Insulators can change their electrical and mechanical properties on being irradiated, although the amount of change naturally depends upon the material of the insulator, and on the type of radiation and its intensity and duration. The general effects are discolouration, evolution of gas, and reduction in resistivity. The resistivity returns to its original value some time, usually hours, after the radiation is removed, and very large doses of radiation are necessary to turn the material into even a poor conductor.

The chemical changes produced by radiation are, however, more serious because they are irreversible. Organic insulators are poorer than inorganic ones in their resistance to radiation, and although the physical effects vary from becoming very hard to turning liquid, permanent chemical changes occur. Rubber, for example, becomes very hard and brittle, and p.t.f.e. turns into powder. P.V.C. evolves hydrochloric acid gas which can corrode the surrounding materials and deform the material or component by pressure. Inorganic insulators suffer from density changes in strong radiation fields, and this may lead to mechanical breakdown owing to dimensional changes in components. Both organic and inorganic materials used for insulation must be carefully selected if the equipment is to work in radiation fields. The selection must be based on consideration of a number of factors such as the chemical composition and the nature, strength and duration of the radiation. The final choice is, necessarily, a compromise.

Wirewound resistors are very stable when irradiated, whereas carbon-film and carbon-composition resistors are much more sensitive. The resistance of carbon resistors is decreased by radiation and changes of up to 15% in value can occur. Metal oxide film resistors are generally very stable, although those with a continuous film of oxide instead of a spiral often change in value by a large amount or fail completely.

Capacitors vary widely in their resistance to radiation, depending upon the nature of the dielectric used. Those which use inorganic dielectric materials such as ceramic, glass and mica are much more stable than those with organic dielectrics such as oil and plastics. Electrolytic capacitors respond in the same way as the organic types. Changes in capacitance occur during radiation, which may be up to 20% in value with oil-impregnated paper capacitors. Apart from the change in capacitance, the evolution of gases from organic dielectrics can cause mechanical failure of the case of the capacitor.

Valves which contain boron in their glass envelopes are prone to failure of the metal-to-glass seal when irradiated, but those with ceramic envelopes are much more stable. Photoelectric cells become less efficient owing to darkening of the glass on exposure to radiation, and the normal photoelectric current is increased by electrons produced by ionization. The characteristics of gas-filled valves are also affected for the same reason.

Semiconductors stand up to radiation fairly well, germanium devices being more stable than silicon. Chemical changes do occur, however, and doped germanium tends to change its properties; n-type tends to become p-type, for example. Transient induced voltages and leakage currents occur during radiation, but they disappear when the radiation is removed.

SUMMARY

Testing of components is essential in order to predict, with any degree of success, the reliability of new equipments or new components, so that errors in design may be minimized. The very high reliability of some modern components makes laboratory testing very expensive, however, for huge numbers of components have to be tested in order to achieve a few failures in a reasonable time.

Thermionic valves have the highest failure rate, fixed capacitors have a fairly high failure rate, followed by semiconductors, transformers, relays, and capacitors. Fixed resistors, plugs and sockets and cables have very low failure rates.

Humidity produces the most severe environmental effects, particularly when associated with high temperature; it lowers insulation resistance and produces spurious voltages. High temperature causes melting, expansion and ageing of components, while low temperature causes contraction, hardening and freezing. The effects of humidity and temperature may be minimized by choosing component materials which withstand them, as well as by drying, cooling or warming where this is possible.

Shock and vibration can be countered by good design of supporting structures, by using anti-vibration mounts and by good positioning of components. Low pressure causes insulation breakdown and oozing of sealed components; the provision of adequate gaps and avoidance of dirt can prevent the former.

Radiation varies in its effects according to its nature, but most damage is done by neutrons and gamma rays, which are very penetrating, and which can change the chemical composition of substances. Insulators are affected more than conductors, and valves more than transistors. Certain component materials withstand radiation much better than others.

CHAPTER 4 FACTORS AFFECTING MAINTENANCE

INTRODUCTION

In Chapters 2 and 3 many aspects of reliability were examined. It was seen that it is not possible, and may not be economically desirable, to obtain very high reliability. Equipment must therefore be maintained, and we will now discuss a number of factors which should be considered when maintenance policy is decided. It is the aim of the technician to ensure that maintenance is carried out with maximum efficiency, and that equipment is operational at the required time and to the required specifications. This aim can be more readily achieved if the reasons for the maintenance policy are understood.

THE FACTORS INVOLVED

The factors which must be considered when the policy of maintenance is worked out are:

Operational requirements
Equipment characteristics
Aids to maintenance
Training
Job environment

There is considerable interaction between these. For example, good aids to maintenance can reduce training costs considerably; the cost of achieving high reliability, which is an equipment characteristic, may be justified if spares are difficult to obtain.

In theory the most efficient system is therefore a certain arrangement of all these factors. In practice this arrangement is impossible to achieve, mainly due to the difficulty of identifying, and making objective measurements of, all the items concerned. Already we have seen some of the difficulties encountered in specifying reliabilities. From the technician's point of view it is important to understand the role of each factor in determining maintenance policy, and some of the major interactions.

OPERATIONAL REQUIREMENTS

The most important factor determining maintenance policy is the operational requirement. By **operational requirement** is meant the function an equipment has to perform and the conditions under which it has to perform its function. It is easy to see that there is a considerable difference between maintaining a satellite transmitter and a walkie-talkie transmitter, or between an airborne radio receiver and a domestic receiver. The difference is not so much one of structure or size, but of operational requirement and relationships between maintenance factors. The satellite must operate reliably, with no maintenance, over a specified period, as must the airborne receiver, perhaps with less stringent requirements. One must be prepared to pay heavily for this by improving reliability. The walkie-talkie or domestic receiver is probably in an environment where spares are readily available, and less than the highest possible reliability can be accepted with its associated lower cost.

It is the object of the maintenance organization to ensure that the operational requirement is met at the minimum cost.

EQUIPMENT CHARACTERISTICS

Equipment characteristics comprise the way the equipment is built electrically and mechanically, and the way in which it works to satisfy the operational requirement. The latter includes such factors as reliability, safety precautions and environment. For example, a particular radio could be briefly described as being a superhet operating on long, medium and short wavelengths, with outputs to a loudspeaker or headphones; these are its **electrical characteristics.** It may also use conventional components on a printed circuit board and be contained in a wooden cabinet 12 in. x 8 in. x 4 in.; these are its **mechanical characteristics.** Finally, it could have continuously variable tuning on each of the three wavebands and a combined on/off switch and gain control; these are the **working characteristics.** If necessary, information about high voltages, correct connection of supplies and use in abnormal temperatures could also be specified as characteristics of the receiver.

Complexity is an equipment characteristic. In the electronic sense this has already been defined as the number of, and interconnections between, components. Clearly the greater the complexity the more difficult is the maintenance task, since it is more difficult to isolate faults or make working adjustments where interconnections are profuse. If the task is difficult then the need for good training, or for aids to the task, increases in importance.

Mechanical structure affects the maintenance task mainly through requirements for manual skills, such as soldering small components, adjusting dust cores, and dismantling and assembly of a receiver tuning cord. All are delicate tasks which have generally to be carried out in confined or otherwise difficult spaces.

From the maintenance point of view reliability is probably the most important characteristic. It has already been discussed at length and it is therefore sufficient to reiterate that the maintenance policy must be determined from a knowledge of the reliability which it is possible to obtain, and the cost of obtaining it. The combination of information about how often equipment goes wrong, with information about how much effort is required to repair it, is an important measurement of operational usefulness.

Extreme cases of each of the factors affecting maintenance are frequently encountered. For example, the very high reliability of fixed cabling is such that maintenance of it is seldom necessary, whereas, at the other end of the scale, there are examples of such gross unreliability that components in equipment have had to be re-designed.

Where an equipment's design leads to conditions which are potentially dangerous for the technician it is impossible to avoid the need for safety precautions. It is also almost impossible to make the precautions foolproof, particularly for technicians who must have access to the components of an equipment when it is working. The need for safety must govern some electrical and mechanical characteristics, and it governs many actions of the maintenance technician. The importance of obeying safety precautions, whether laid down in instructions or learned in basic training, can never be over-emphasized.

The final characteristic we will mention is the equipment's working environment, i.e. the conditions external to the equipment in which it must operate. In discussing reliability the effect of environment was considered, and it will therefore be appreciated that there is a close link between environment, reliability and maintenance policy. As examples, submarine cables must be highly reliable because they are required to operate in an inaccessible environment. A car electrical system is usually regarded as being not highly reliable under the rigours of winter motoring, but the availability of spares and accessibility of the components makes this acceptable.

In spite of the fact that the equipment characteristics which are desirable for efficient maintenance are well known, it is often the case in practice that pressure to meet a certain operational requirement is such that the best maintenance backing is not provided. It may then be possible to improve matters by adjusting other maintenance factors; if not, a deterioration in operational performance must be accepted.

AIDS TO MAINTENANCE

Aids to maintenance are the tools, test equipment and information which are not required for an equipment's normal operation, but which are desirable for carrying out maintenance.

Some aids to maintenance are always available, but the amount is extremely variable. At one end of the scale, a circuit diagram may be all that is available; at the other end, automatic test equipment may be installed. The type of aid available primarily influences training and the ability level of the men who are required to carry out maintenance. If a circuit diagram is the only aid provided, technicians must be trained to use the information in the diagram for such matters as determining the normal voltages to be expected, and for making decisions about test procedures. The men selected for training to this level of skill must therefore be of high ability. On the other hand, if the task is thoroughly documented, and the technician has only to obey instructions, little training is required and men of lesser aptitude can be selected. It is important to note that where maintenance tasks are automated, the technician's job often only changes from maintaining the equipment to maintaining the automating devices, and men of high ability are therefore still required.

Aids to maintenance, and how they can be improved, form the subject matter of a large section of this book and details are given in later chapters.

TRAINING

It is seldom that men can be found who, without training, satisfy the requirements of a particular maintenance task. Because training takes a lot of time and money it is one of the most important factors in determining maintenance policy. The training requirement can be summarized as the difference between the ability required to do a task and the initial ability of men selected for it. That is: initial ability plus training gives the required ability.

It is possible to reduce the costs of training either by raising the selection standards for technicians and shortening the training course, or by improving the aids to maintenance which are intended to simplify the task and reduce the required ability. Chapter 11 deals more fully with training for maintenance.

JOB ENVIRONMENT

The conditions in which a technician works are as important as the conditions in which equipment is operated. Apart from the physical comfort of the working space, other factors have to be considered, such as the availability of spares, the amount of supervision and guidance given, the time available to complete tasks and safety precautions.

There are two aspects to the provisioning of spares: the cost of the hardware, and the organization of the supply system. Generally speaking, where the cost of producing a unit is small, as is the case with printed circuits, the maintenance policy will provide for complete replacement of unserviceable units. In cases where the spare unit is expensive, it could well be more economical to increase the reliability, or even to change the design of the equipment so that it is easier to maintain. Supply of spares is a large subject which is often treated mathematically in management texts.

From the point of view of maintenance, the provision of spares affects the **availability** of the equipment. It will be remembered that availability was defined in Chapter 2 as

$$\text{Availability} = \frac{\text{m.t.b.f.}}{\text{m.t.b.f.} + \text{m.t.t.r.}}$$

where m.t.b.f. is the mean time between failures, and m.t.t.r. is the mean time to repair. The m.t.t.r. here includes not only the actual time spent in repairing a fault, but also the time waiting for spares; thus it represents the total time that the equipment is out of commission. A good spares organization will reduce the time waiting for spares to a minimum, and will thus improve the availability of the equipment – and this is what interests the user most.

MAINTENANCE POLICY

The best maintenance policy is obtained as a result of the optimum combination of the contributing factors which have been mentioned, although it is scarcely possible to do this exercize mathematically. It is sufficient for the technician to appreciate that the maintenance policy he is carrying out is the result of a balance between these factors.

It is also worth bearing in mind that policy decisions can be altered as more accurate information becomes available. When equipment has been in use for some time, accurate information is obtainable concerning all the factors which were predicted in the development stages. This information can be collected and used to modify the maintenance policy and increase efficiency. The technician is a key person in this feedback process, and it cannot be over-emphasized that the meticulous recording of maintenance data is an important part of a technician's task, although, in all too frequent cases, he is not told of the reasons for the collection of this data. Examples of data are recording the completion of routine tasks, recording details of operation, and recording faults. The last is of particular importance, both as specific information about an equipment, and as general information about components. With the generally high m.t.b.f.'s which are now obtainable, a long time elapses before sufficient data can be obtained to check reliability predictions and the opportunity to collect valuable data must not be wasted by defects in recording.

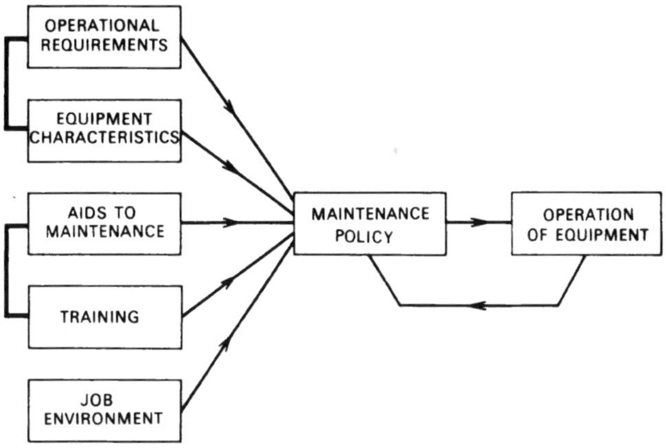

Figure 4.1 Factors affecting maintenance

Figure 4.1 illustrates the factors affecting maintenance which have been discussed in this chapter, and the heavy lines denote the most important interactions between them.

SUMMARY

The factors which must be taken into account when maintenance policy is decided are operational requirements, equipment characteristics, aids, training and job environment.

Operational requirement refers to the function of the equipment and the conditions under which it has to operate.

Equipment characteristics refer to its electrical and mechanical construction and properties, and the way in which it works.

Aids to maintenance consist of the tools, test equipment and information necessary.

Training for maintenance covers the teaching that is required to turn unknowledgeable or unskilled people into competent technicians.

Job environment, or the conditions under which a technician works, includes not only physical environment but spares provision, supervision, time and safety.

CHAPTER 5 PREVENTIVE AND CORRECTIVE MAINTENANCE

INTRODUCTION

Maintenance has two aims. The first is to prevent failure by routine checks on units or components, for which a predictable rate of wear occurs, with the object of replacing any that are nearing the end of their life. This is called **preventive maintenance.** The second is the detection, location and repair of faults when and as they occur. This is called **corrective maintenance.** For most equipment both types of maintenance are carried out.

It should be appreciated that in any one piece of equipment some components will be in their wear-out period and be subject to preventive maintenance, whereas other components will still be in their useful life period and only fail on a random basis; these latter components will be subject to corrective maintenance in the case of random failure. Preventive maintenance is thus intended to provide checks to cover the wear-out period of a component's life, including detecting the onset of this period, as well as replacing worn-out parts. Corrective maintenance is intended to cover the useful life period of a component or equipment. These periods were dealt with in Chapter 2. In electronic equipment valves may be subject to preventive maintenance by regular checks on their cathode emission, whereas resistors in the same circuit may only be replaced as they fail.

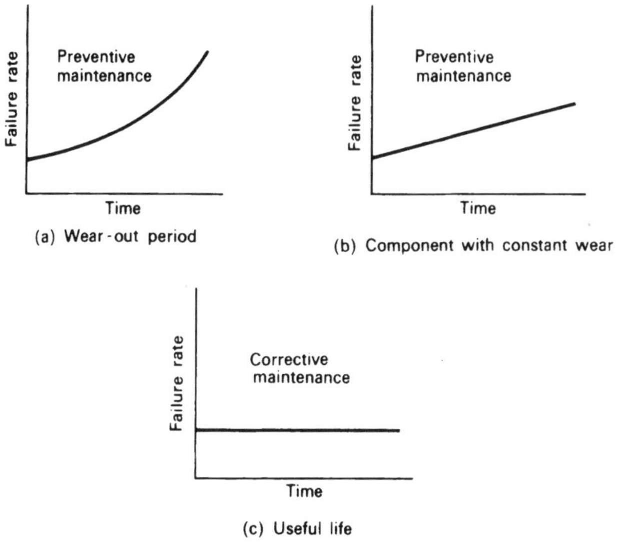

Figure 5.1 Preventive and corrective maintenance

The difference between the two categories of maintenance is related to the different probabilities of failure as shown in Figure 5.1. Figure 5.1 (a) shows the probability of failure, or failure rate, against time during the wear-out period; Figure 5.1 (b) shows the failure rate against time for a component which has a constant rate of wear; these are the conditions in which preventive maintenance is required. Figure 5.1 (c) shows the failure rate, against time, during the useful life. As we have seen already, this is a period of random but constant failure for which corrective maintenance is appropriate.

Most of this chapter is concerned with corrective maintenance. The concepts of fault detection, fault location and fault rectification are developed, and methods of fault location, and of checking for faults, are discussed in some detail.

PREVENTIVE MAINTENANCE

Preventive maintenance increases reliability by predicting failures and preventing them from occurring. This should not be confused with the steps taken to improve reliability in the design stage which require no subsequent action by the maintenance technician.

When a component or unit has a known rate of wear the task is either one of measuring directly the amount of wear, for example the length of the carbon brushes in a dynamo, or of keeping an account of wear in terms of another quantity related to it. Examples are the number of operations of a switch, the time an equipment is operating and the varying value of the current gain of a transistor. Such wear leads to the need for routine maintenance schedules, such that failure to carry these out and take any necessary remedial action would result in certain failure of the equipment.

Another reason for preventive maintenance is the need to check that parts of the equipment are working under optimum conditions. It may be the case that, although according to its outputs an equipment is working perfectly satisfactorily, some components are not working under optimum conditions. The power distribution between components may not be that required by the design, or a valve may be working on an inefficient portion of its characteristic, for example. This aspect, however, really falls into one of the two maintenance categories being discussed. A consistent variation in the characteristics of a component should be treated under the heading of preventive maintenance, whereas there is no point in attempting to rectify random variations until they are large enough to result in a fault and be treated as corrective maintenance.

When the failure of a component or unit is random, i.e. the probability of failure at any specific time in its life is a constant, there is no necessity for preventive maintenance. Introducing preventive measures for random failures can only lead to a reduction in equipment reliability,

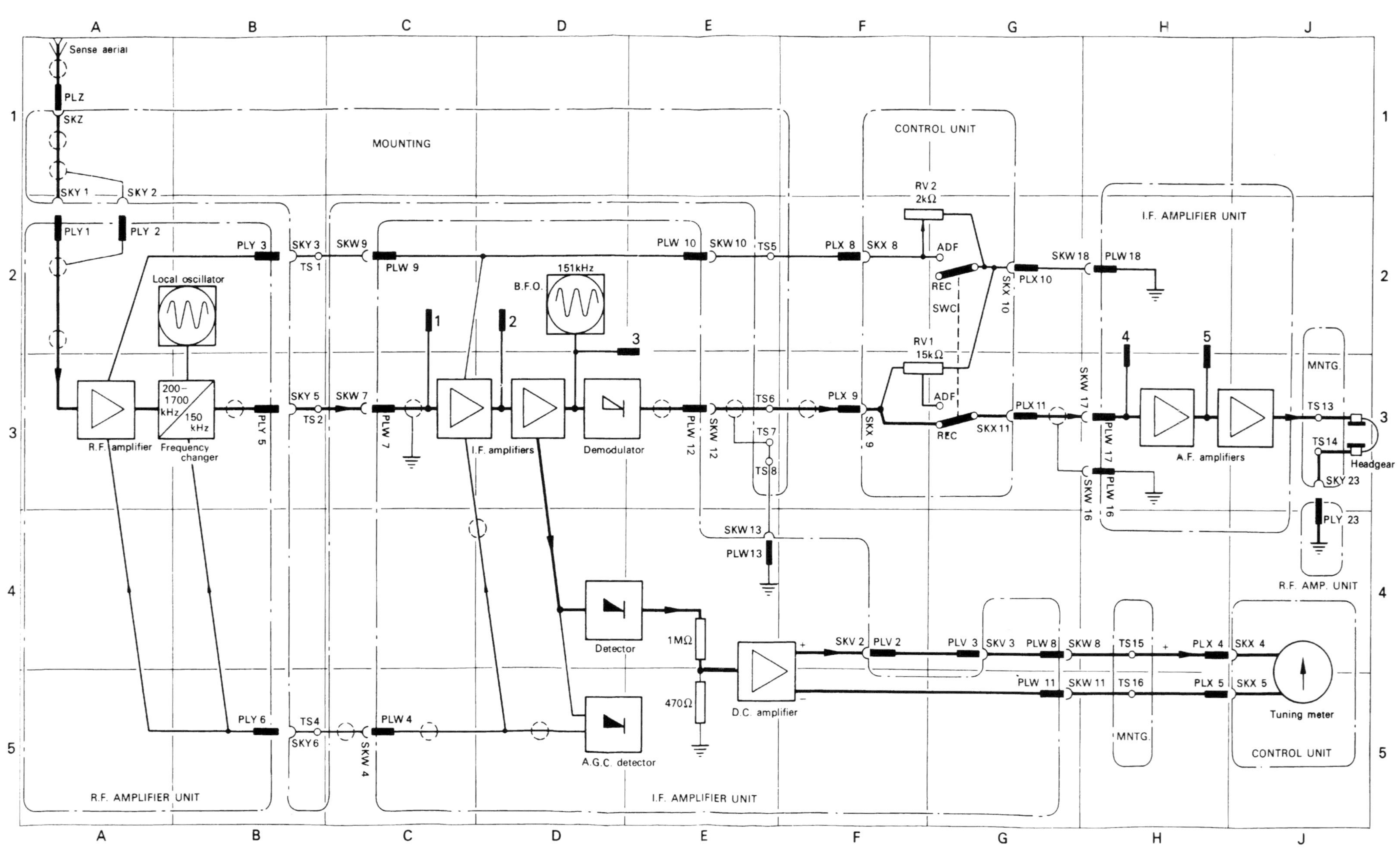

Figure 5.2 An airborne receiver

because, in general, the more the normal operation of an equipment is interfered with, the less reliable it becomes. It used to be thought that regular checking of an equipment would detect potential failures, and hence improve overall reliability, but this is certainly not so in the case of random failures. No routine checking can predict random failures.

THE FUNCTIONAL DIAGRAM

In order to illustrate and explain the logical procedures advocated in this book, the functional diagram shown in Figure 5.2 will be used extensively. It is not possible to use a diagram of great size and complexity because of the difficulties involved in incorporating one in a book, and Figure 5.2 represents a fairly straightforward piece of electronic equipment. The principles to be developed are, however, applicable to the most complex equipment, and the benefits of using and applying them become more spectacular as the equipment becomes more complex.

Figure 5.2 represents a communication receiver which is part of an airborne automatic direction finder. It will be noticed that very few components are shown on the diagram, which is intended to illustrate the function of the equipment by splitting it into its interrelated stages, and showing the main paths of the received signal through them. The method of drawing such diagrams, and the rules for dealing with such matters as alternative paths and feedback loops will be described in detail in Chapter 8.

The diagram is read from left to right, and in this case we start at the aerial in square A1. The screened aerial lead terminates in a plug PLZ which fits socket SKZ in the mounting. The signal continues along the screened lead from SKZ to socket SKY1 in the same mounting, and thence to plug PLY1 on the r.f. amplifier unit. The mounting is a suspension rack on anti-vibration mounts, and the r.f. amplifier unit fits into it with plug and socket connections. As is common in electronic equipment, other sockets giving connections to other parts of the equipment are also located in the mounting. Plugs and sockets such as SKY2 and PLY2 provide bonding and earthing of the outer screening conductors.

The signal path follows the heavy flow line to the r.f. amplifier and frequency changer and on to the first i.f. amplifier (C3). The chain lines, such as in A5, represent the physical boundaries of the various removable units such as the mounting, r.f. amplifier unit, i.f. amplifier unit and so on. Since the diagram concentrates on the signal flow, and is not concerned with the physical location of units or components, the boundaries are bent as necessary. Here it happens that the boundary of the r.f. amplifier unit is straightforward except for a detached boundary in J4, but that of the mounting is drawn to accommodate its plugs and sockets in their correct functional positions.

The main signal flow line is thick, with arrows pointing to the right, but two other thinner lines join the r.f. amplifier stage (A3). One is an a.g.c. line from the a.g.c. diode (D5), and the other is a gain control connection which is switched in, through the use of the variable potentiometer RV2 (F2), when the switch C in G2 is in the REC position. Feedback paths, such as the a.g.c. line, are shown with arrows pointing to the left; the gain control does not represent a signal or feedback path.

Test points are shown as test sockets (e.g. TS1 in B2, TS2 in B3) or as test points (e.g. 1 in C2). In this equipment test sockets are all brought out to the front of the unit and are easily accessible for voltage or waveform measurements. The test points are plugs, also brought out to the front panels, which are mainly designed for signal injection from a special test signal generator.

After passing through the i.f. amplifiers the signal reaches the demodulator stage (D3). For heterodyned c.w. reception a beat frequency oscillator (D2) can also be switched in. Other outputs from the last i.f. amplifier are taken to a separate a.g.c. diode stage (D5), and to a stage (D4) which drives a tuning meter. The latter stage also provides an input to a series of amplifiers which eventually drive a motor connected to a goniometer search coil; these stages are not shown on the diagram.

The output from the demodulator passes to the control unit, which contains the switches and controls for the whole equipment, through SKX9 to switch C (G2 and G3). In the REC (receive) position this switch connects the output from the demodulator directly to the first a.f. amplifier (H3), and the variable resistor RV2 (F2) acts as an r.f. gain control. Connections in the receiver are such that the automatic direction finding circuits are made inoperative when switch C is in the REC position. In the ADF (automatic direction finding) position switch C short circuits the r.f. gain control RV2 but introduces RV1 (F3) as an a.f. gain control.

Finally, the signal is amplified by two a.f. amplifiers in cascade (H3 and J3) and applied to headphones (J3). The a.f. stages are physically contained in the removable i.f. amplifier unit, and the container is labelled as such (H2).

CORRECTIVE MAINTENANCE

The conditions under which corrective maintenance is needed have already been mentioned, but apart from random failures, it is clear that if any equipment is allowed to operate without preventive maintenance it will eventually fail. Lack of preventive maintenance may be due to many causes, not all of them avoidable; operational necessity or lack of information concerning wear, for example. But in both circumstances, random failures and wear-out failures, the technician has three phases in his corrective maintenance task:

1. He must detect the fact that a failure has occurred: this is **fault detection.** Although this aspect may at first sight seem trivial, it is not always immediately apparent that a failure has in fact occurred, as will be discussed below.
2. He must find the faulty component; this is **fault location.**
3. He must repair or change the faulty component; this is **fault rectification.**

These three phases of fault-finding and repair are very important in the maintenance task. New methods of improving fault location in particular will be gone into in detail.

FAULT DETECTION

The report about a failure seldom provides the technician with much information; indeed he may not be certain that the failure is caused by a fault, and is not the result of incorrect operation. The first requirement is therefore to carry out a functional check, i.e. to test the operation of the equipment. This will provide both confirmation that a fault does exist, and information to assist in diagnosis. For example, if a television receiver is reported as showing no picture, the first check is to see that the brilliance control is correctly set. If there is still no picture, check is made to see if the alternative vision channels give an output, or if the sound channel is operating. These are functional checks. Similarly, referring to Figure 5.2, a fault reported as no sound provides little information for fault location, whereas a systematic check of the equipment, including the meter output and sound using the alternative gain control system, provides information which considerably narrows down the faulty area.

A functional check is clearly most important in the case where an equipment is reported faulty but little indication is given of symptoms.

It often happens that an enthusiastic technician, finding a familiar fault symptom, ignores the functional check and commences his checks in a specific area on the basis of that symptom alone. This is a technique which sometimes pays dividends but there is no doubt that in the long run, and particularly with complex equipment, it is an inefficient and thus undesirable method. A systematic collection of information is the only correct procedure.

FAULT LOCATION

Fault location, particularly in complex equipment, is the technician's most difficult task. The reason for this is that it requires decisions to be made, usually based on a large amount of data, for which very little guidance has hitherto been given. The decisions may require the recall of some of this information from previous checks, and a considerable memory load may be placed on the technician. One of the objects of this book is to simplify the task of fault location by providing guidance on how best to go about the task, hence a large proportion of it is devoted to fault location.

METHODS OF FAULT LOCATION

Given an equipment which is known to contain a fault , there are a number of methods of determining which is the faulty component; each has advantages and disadvantages which we will go into in detail. Consider first the complete pattern of possible methods of fault location by reference to Figure 5.3.

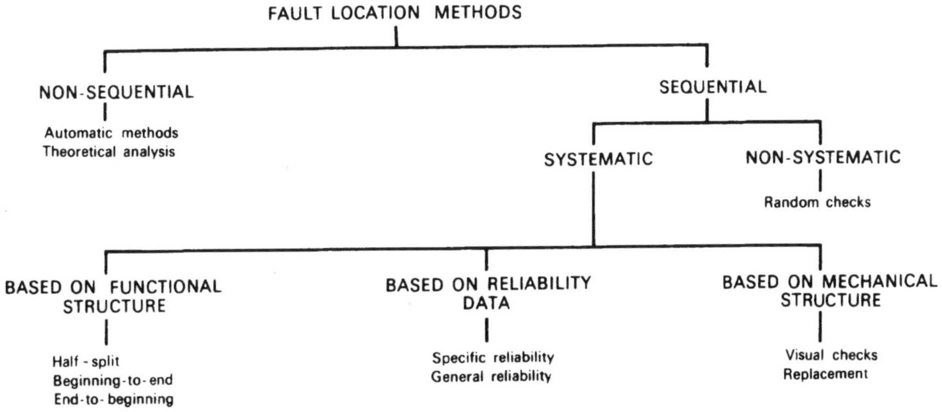

Figure 5.3 Methods of fault location

NON-SEQUENTIAL METHODS OF FAULT LOCATION

The first division is made between methods which require the technician to make a series of checks or measurements to narrow down the faulty area, and methods in which the fault is located by one measurement or a number of measurements made rapidly and automatically. These are called **sequential** and **non-sequential** methods respectively. The latter method does not require the technician to decide on a series of checks in order gradually to narrow down the faulty area. It is seldom used by technicians but an increasing interest is being taken in it as an automatic method.

An example of a non-sequential method is the determination of a fault by the measurement of transfer functions, by which is meant an examination of the relationships between input and output signals. This example applies primarily at circuit level. A signal of varying frequency is applied to the input of a circuit and the output characteristic curve is measured. This output is compared with a set of pre-determined characteristics obtained with particular component failures in the circuit. The closest match will then indicate the fault. This technique has limited application, but it has been shown to be feasible for small linear electronic circuits.

In a circuit tuned to a particular frequency a fault in any of the tuning components would change the resonant frequency, and the nature of the detected change can be used to determine which component is faulty. Similarly, resistor faults, whilst not altering the resonant frequency, could change the circuit gain at other frequencies; this gain change can also be related to the failure of a specific component.

A similar method which compares voltages, instead of frequency characteristics, is to measure voltages at a number of key points in a circuit and compare these with pre-determined sets of voltages obtained for particular component failures.

Both methods require that a set of measurements be taken from the equipment and then a large number of comparisons made. Because of the large number of comparisons required for a circuit of any size, the methods lend themselves best to maintenance using computers or other data processing equipment.

Some attention has also been paid to a method of diagnosis which requires no direct connection to the faulty circuit. The method involves the detection of some secondary phenomena which would indicate the serviceability of components. Examples of such phenomena are: X-ray absorption, in which circuit defects can be detected in the same manner that defects in the human body are detected using X-rays; infra-red radiation measurement, in which a thermal characteristic of a circuit can be obtained; r.f. radiation detection; and the measurement of magnetic or electrostatic fields. For some phenomena this method is similar to the other two examples, in that a set of measurements has to be compared with a large number of pre-determined characteristics related to specific component failures. For others the defect can be readily detected from one measurement e.g. an open circuit wire detected by X-rays.

A more commonly encountered non-sequential method is theoretical analysis. Here an attempt is made to deduce, from first principles, the component or components and type of failure which could cause the detected symptoms. For example, with a circuit which produces a sine wave output, it might be possible to note a particular distortion in the waveform and attribute this to a certain component failure in the circuit. This is an impressive and sometimes effective technique, but it is also a difficult one which is more suited to the designer than the technician.

SEQUENTIAL METHODS OF FAULT LOCATION

Generally the technician will not meet the non-sequential methods described, except when they are incorporated in automatic devices. It is more likely that he will have to use techniques which require a series of checks to be made in order to narrow down the faulty area i.e. he will have to use sequential testing. The series of checks can be described as either **systematic** or **non-systematic.** It is systematic if a set of pre-determined rules is used to decide on each check; it is non-systematic if, although each check is made in the faulty area, no particular principles or rules are used in deciding which checks to make. As an example of a non-systematic series, refer to Figure 5.2. A first check could be at test point 2, and, if this showed a correct output at this test point, the next could be at test point 6; then test point 3 if 6 proved incorrect. The faulty area is thus reduced by a series of checks, but the order of the checks in this case is not determined by any rule.

It is much more likely that the eliminating process will be governed by a set of rules. The most important rules which are used, sometimes perhaps sub-consciously, and their relative merits, will be discussed in detail.

SYSTEMATIC CHECKING

Referring to Figure 5.3, three broad divisions are made within the field of systematic checking.

First, decisions concerning where to check can be based on reliability data. The reliabilities of different components are fairly well established and, if the equipment is similar to one in which the components are of known reliability, checks and procedures can be made on the basis of these component reliabilities. For example, a decision may be made to change all the valves as an early check in a fault location procedure. It may be that reliabilities of the specific components within an equipment itself, instead of in a similar equipment, are available, and these would provide a better basis for decisions. This concept applies at any level; e.g. information may be available concerning the relative reliabilities of removable units rather than components. In the equipment of Figure 5.2 it may be the case that, of the last 100 failures, 80 have been in the i.f. amplifier, 15 in the r.f. amplifier, 5 in the control unit and none in the mounting. On a reliability basis the units would, therefore, be changed in the order: i.f. amplifier, r.f. amplifier, control unit and mounting.

Either general or specific reliability data can, therefore, form the basis for decisions and on this basis unreliable components or units would receive the earliest attention.

Secondly, the choice of a sequence of checks could be based on the **functional structure** of an equipment. By functional structure is meant the relationships between the component parts as far as the operation of the equipment is concerned; this information is provided in the form of block diagrams and circuit diagrams. A large number of methods could be included under this heading, but the three most common ones are: commencing with a check near the input of an equipment and eliminating areas by checks which gradually move towards the output; the reverse of this, i.e., commencing near the output and working towards the input; and a method known as the **half-split**, in which each check is made as close as possible to the centre of the faulty area. The half-split can be shown to be the most efficient method, and it is discussed in detail in Chapter 6.

The input-to-output method applied to Figure 5.2 would commence with a check at test point 1, and proceed with checks at test points 2, 3, 4, etc., until the fault is located. Output-to-input would commence with a check at test point 6 and work towards test point 1. Checks based solely on the half-split method would commence at or near test point 4 and further checks would depend on the result obtained at this point.

The third method of systematic elimination is based purely on the mechanical structure of the equipment. Looking for the fault in the purely visual sense comes under this heading, providing it is done systematically. For example, a search can be made by looking for the fault at one side of a chassis and looking at each component and connection in turn, working toward the other side of the chassis. A fault may either be seen directly, for example a broken wire or shorted terminal tags, or be detected by some secondary effect such as melting wax or burnt paint. The care necessary in interpreting the latter symptom has already been mentioned in the chapters on reliability; the fact that a resistor has overheated does not always mean that the resistor itself is faulty.

Changing units or components in turn, although not in an order based on reliability, is also a technique which is based purely on mechanical structure. An example of this method is an equipment constructed of printed circuit cards mounted in rows, in which a faulty card is located by changing each in turn moving along the rows.

CHOOSING THE BEST METHOD OF FAULT LOCATION

Fault location is carried out using a combination of the above methods, but usually one method predominates. The particular combination of methods used will depend on factors such as, whether reliability information is available or not, whether units can be easily removed or not, and the availability of test points. A method based on reliability is clearly useless when there is a lack of failure rate data, and a method based on unit replacement is not possible for an equipment which is not constructed with easily removable units.

Whatever the combination of methods used, most maintenance tasks require the technician to make a series of checks in order to reduce progressively the faulty area. If maintenance efficiency is related to the number of checks which are required to locate the fault, so that the fewer the number of checks made, the better is the technician's performance, then the best method to use must be the one which uses the lowest mean number of checks. This usually results in the shortest time to locate the fault. It is important to note that the mean number of checks must be considered, and also that the mean must be taken from a large number of maintenance tasks. The reason for this is that for one specific task a particular method could be highly successful, but in a large number of tasks it could be much less so. Selecting a component at random for checking is an example. If the first component selected proves to be the faulty one the method is highly successful in this instance, but a large number of maintenance operations would show that, in fact, such a random method is generally very inefficient.

It is possible to calculate the mean number of checks which are required for different methods of fault location, and thus to compare their effectiveness. This has been done for three cases: non-systematic checks which are purely random; input-to-output checking, which is effectively the same as output-to-input when considering the number of checks required; and the half-split method. In each case a linear chain of components is considered, in which it is known that one component is faulty. The formula for each method is:

(a) Random checks $c = \Sigma t.P$

(b) Input to Output $c = \frac{1}{2n}(n-1)(n+2)$

(c) Half-split $c = 3.32 \log n$

In these formulae, c is the mean number of checks required to locate the faulty component, n is the number of components which could be faulty, t is the number of checks in a particular solution and P is the probability that the solution requiring t checks will be used. Log refers to logarithm to base 10.

The proofs of the formulae (a) and (b) are given in the Appendix to this chapter, and that for (c) is given in Chapter 6.

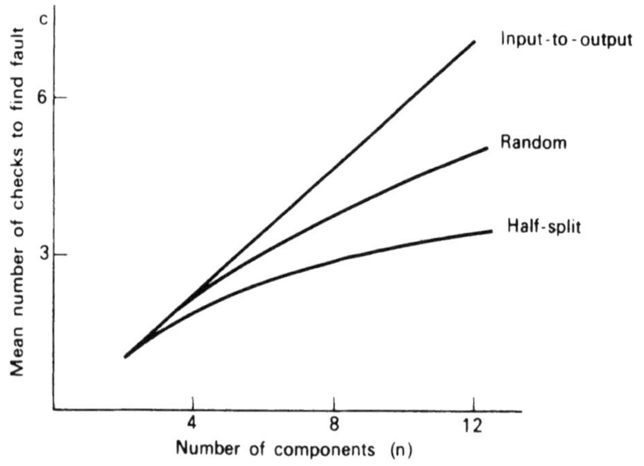

Figure 5.4 Comparison of methods of fault location

A graph which compares these three results is shown in Figure 5.4. It should be noted that, although the graphs are drawn as continuous lines, the variables can only take discrete values. It is not possible to have half a component or half a check.

It can be seen that for any value of n, no method betters the half-split, and that for large values of n the half-split is much superior to the other methods. Although, for convenience of calculation, a linear chain of components has been taken, each method is applicable to more complex arrangements. The final result, however, is the same. The overall result is not surprising when it is realized that the half-split method is a fundamental one which is applicable to all searching tasks.

THE RELIABILITY METHOD

Now let us consider how a method of fault finding based on reliability would compare with the above methods. It is not possible to represent this method on the graph of Figure 5.4 because the mean number of checks will depend on the relative reliabilities of the components in the chain. Nevertheless, it is possible to examine some extreme cases to obtain a broad picture. Suppose that we have a chain of four components, and we know that one of them is faulty. Let us also give the components m.t.b.f.'s of m_1, m_2, m_3, and m_4. Thus in a given time, say 10,000 hours, the expected number of failures, N, of each component is

$N_1 = \frac{10{,}000}{m_1}$, $N_2 = \frac{10{,}000}{m_2}$, $N_3 = \frac{10{,}000}{m_3}$,

and $N_4 = \frac{10{,}000}{m_4}$. The total number of failures is, therefore, $N = N_1 + N_2 + N_3 + N_4$. The method we are interested in is checking in order of reliability, starting with the least reliable component, (i.e. that having the lowest m.t.b.f. and hence the largest number of failures in a given time), and working towards the most reliable. Let us assume that this order in our example is m_1, m_2, m_3, m_4, or, in terms of failures, N_1, N_2, N_3, N_4. In the 10,000-hour period being considered, the least reliable component can be expected to fail N_1 times, and hence, since it is the first component to be checked, will be subject to $(N_1 \times 1)$ checks. The next reliable component is expected to fail N_2 times, but since it is always the second to be checked, $(N_2 \times 2)$ checks will be carried out in order to test it each time. Similarly, the other two components will result in $(N_3 \times 3)$ and $(N_4 \times 4)$ checks being carried out. Thus the total number of checks which can be expected to be necessary in 10,000 hours is

$$(N_1 \times 1) + (N_2 \times 2) + (N_3 \times 3) + (N_4 \times 4)$$
$$= N_1 + 2N_2 + 3N_3 + 4N_4$$

This number of checks will reveal all faults occurring i.e. a total of N faults. It is assumed here that, after checking three components and finding them serviceable, it is still worth checking the fourth; it is also assumed that two or more components do not fail simultaneously. The mean number of checks required to find one fault will, therefore, be

$$c = \frac{1}{N}(N_1 + 2N_2 + 3N_3 + 4N_4).$$

For convenience of calculation let us take N as 100 total failures in 10,000 hours. We can now examine the effect on c of various assumed reliability distributions, as follows:

(a) If $N_1 = N_2 = N_3 = N_4 = 25$ then $c = 2.5$

(b) If $N_1 = 30, N_2 = N_3 = N_4 = \frac{70}{3}$ then $c = 2.4$

(c) If $N_1 = 40, N_2 = 30, N_3 = 20, N_4 = 10$ then $c = 2.0$

(d) If $N_1 = 80, N_2 = N_3 = N_4 = \frac{20}{3}$ then $c = 1.4$

Comparing the values obtained for c with the value 2.0 which is obtained if the half-split method is used with a chain of 4 components in series, we can conclude:

(a) Whether or not checking components in an order dependent on reliability is worthwhile is governed by the relative component reliabilities.

(b) To be more efficient than the half-split method there must be a very large difference in component reliabilities.

Since accurate reliability figures are not easily obtainable, for the reasons given in Chapter 2, and components which prove particularly unreliable are usually modified, checking by this method is not recommended except in extreme cases.

FAULT RECTIFICATION

Having located the defective unit or component the technician's task is to repair or replace it. The choice of whether to repair or replace should have been made when the maintenance policy was decided. Similarly the decision whether to repair a unit, or to return it to a central depot for repair, should also have been made.

The skills involved in repair are generally simpler than those required for fault location, and there are many books dealing with such matters as soldering and the use of hand tools. Although they form an extremely important aspect of the maintenance task, they are not, therefore, dealt with in this book; however, certain aspects relating to the use of test equipment are covered in Chapter 10.

It is possible that repair action itself may result in faults being inserted in the equipment. This clearly must be avoided if maintenance costs are to be kept to a minimum:

THE COMPLETE MAINTENANCE CYCLE

Having separately considered the various elements associated with corrective maintenance, we are now in a position to piece together the complete corrective maintenance cycle. This is done by reference to Figure 5.5.

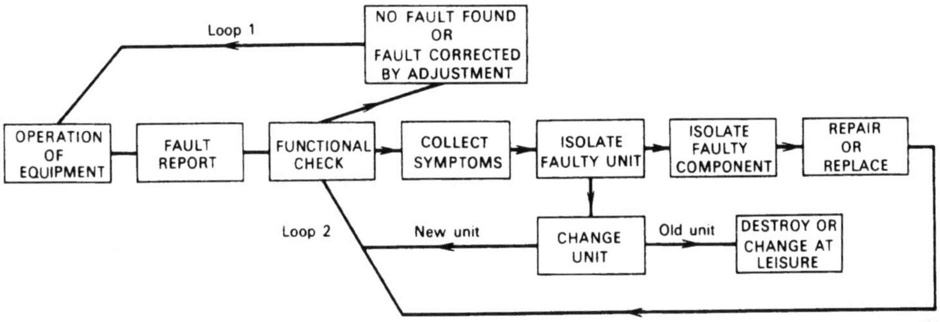

Figure 5.5 The corrective maintenance cycle

Fault detection is represented by the functional check, fault location by isolation of the faulty unit or component, and fault rectification by the repair or replace alternatives shown.

It is important to note the feedback loops. Loop 1 represents the fact that some reported faults are not confirmed by the technician. The reasons for this might be operator error, or the occurrence of an intermittent fault. Loop 2 represents the fact that repaired equipment has to undergo a functional check before it is returned to normal operation. This functional check serves as confirmation that the repair has been effectively carried out, and that in carrying out the repair no other faults have been introduced into the equipment.

THE IMPORTANCE OF OBSERVATION

One aspect of maintenance which has been mentioned as a method of location but which covers all elements, is the importance of observation. It is sometimes the case that failures can be located by direct observation. This is of greatest importance in the early stages of fault location, where the saving of time is greatest if remedial action can be taken before starting a narrowing-down procedure.

A word of warning is, however, necessary.

Many observed symptoms are indirect ones which are remote from the true fault. A burnt-out resistor is often caused by a fault elsewhere in the equipment, and a blown fuse seldom indicates a faulty fuse!

SUMMARY

Preventive maintenance is intended to prevent failure by carrying out routine checks and replacing components that are approaching their wear-out period.

Corrective maintenance is the detection, location and repair of random faults.

Fault detection involves checking that a fault has occurred, and that the failure is not due to incorrect operation; functional checks are used for this purpose.

Fault location is the most difficult task for the technician, and methods which simplify and speed up fault location are extremely desirable. The first division is into **non-sequential** and **sequential** methods. In the former a number of measurements are made, usually automatically, and comparison is made with a set of pre-determined characteristics. In the latter a series of checks is made which gradually narrows down the faulty area. Non-sequential methods are suited mainly to automatic testing.

Sequential methods can be further divided into **systematic** and **non-systematic**, of which systematic methods of fault location are to be greatly preferred.

Within the field of systematic fault location, methods can be based on **reliability data**, on the **functional structure** of the equipment and on the **mechanical structure.** In general a combination of these three methods is used.

Fault location using reliability data can sometimes be effective, but the technician usually has to make a series of checks to localize a fault. Although attention to mechanical structure, by noting visual defects, can assist, the results of visual inspection have to be treated with caution. The most powerful method of fault location is by systematically checking the electrical structure.

The electrical structure can be checked by working from input to output, or vice versa, or by using the **half-split** method, to be described in detail in Chapter 6. The half-split method always results in taking the least number of checks to find a fault.

APPENDIX TO CHAPTER 5

METHODS OF FAULT LOCATION

In the two methods analyzed we will consider **n** components connected in series, one of which is known to be faulty. It will be assumed that all components have the same m.t.b.f. or probability of failure. The conditions are therefore as shown in Figure 5.6.

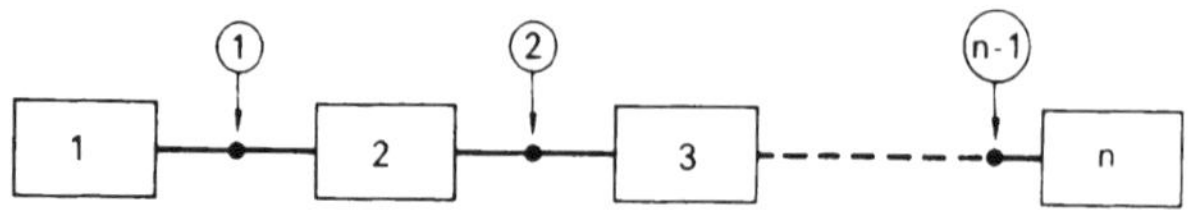

Figure 5.6 Components in series

RANDOM CHECKING

In this method the selection of a test point is random but all tests are made in an area known to be faulty.

The probability that any one component is faulty, since there are n components of which one is known to be faulty, is

$$\frac{1}{n}$$

The probability of selecting any one of the (n–1) test points is

$$\frac{1}{n-1}$$

The probability of selecting a test point 'a' components from the input end is therefore

$$\frac{1}{n-1}$$

The probability that the fault is in one of the 'a' components is

$$\frac{a}{n}$$

and the probability of selecting the particular test point AND of the fault being in one of the 'a' components is thus

$$\frac{a}{n\,(n-1)}$$

Similarly, the probability of selecting a test point 'a' components from the output AND of the fault being in one of the 'a' components is

$$\frac{a}{n\,(n-1)}$$

The probability of narrowing down the faulty area down to 'a' components is therefore the sum of the probabilities of narrowing the faulty area down to 'a' components from the input end and of narrowing the faulty area down to 'a' components from the output end. This is

$$\frac{2a}{n\,(n-1)}$$

Any sequence of checks narrows the area down to 'a', then 'b', then 'c', then 'd', etc., components. The probability that a particular sequence will be used is therefore

$$P = \frac{2a}{n\,(n-1)} \cdot \frac{2b}{a\,(a-1)} \cdot \frac{2c}{b\,(b-1)} \cdot \frac{2d}{c\,(c-1)} \cdots\cdots$$

$$= \frac{2}{n\,(n-1)} \cdot \frac{2}{(a-1)} \cdot \frac{2}{(b-1)} \cdot \frac{2}{(c-1)} \cdots\cdots\cdot$$

If t tests are needed to reach the faulty component then there will be t terms in the expression. The probability, P, of reaching the fault in t tests is therefore:

$$P = \frac{2^t}{n\,(n-1)} \cdot \frac{1}{(a-1)} \cdot \frac{1}{(b-1)} \cdot \frac{1}{(c-1)} \quad \ldots \text{(t terms)}$$

The numbers a, b, c, etc., may have any value from 2 to (n–1), but they must all have different values because each check is made at a different test point. (a–1), (b–1), (c–1), etc., are therefore the numbers from 1 to (n–2), and there are (t–1) of them.

For this method of testing the mean number of checks in a large sample will be $c = \Sigma t\,P$, where t takes the value of each of the integers between 1 and (n–1), and P is obtained from the above formula.

As an example consider the case of n = 6; t can have any value from 1 to 5.

When t = 1, $P_1 = \frac{2}{6.5} = \frac{2}{30}$ $\qquad tP_1 = \frac{1}{15}$

When t = 2, $P_2 = \frac{4}{6.5}\left(\frac{1}{4} + \frac{1}{3} + \frac{1}{2} + 1\right)$

$= \frac{4}{30} \cdot \frac{50}{24} = \frac{5}{18}$ $\qquad tP_2 = \frac{5}{9}$

When t = 3, $P_3 = \frac{8}{6.5}\left(\frac{1}{4.3} + \frac{1}{4.2} + \frac{1}{4.1} + \frac{1}{3.2} + \frac{1}{3.1} + \frac{1}{2.1}\right)$

$= \frac{8}{30}\left(\frac{1}{12} + \frac{1}{8} + \frac{1}{4} + \frac{1}{6} + \frac{1}{3} + \frac{1}{2}\right)$

$= \frac{8}{30} \cdot \frac{35}{24} = \frac{7}{18}$ $\qquad tP_3 = \frac{7}{6}$

When t = 4, $P_4 = \frac{16}{6.5}\left(\frac{1}{4.3.2} + \frac{1}{4.3.1} + \frac{1}{4.2.1} + \frac{1}{3.2.1}\right)$

$= \frac{16}{30}\left(\frac{1}{24} + \frac{1}{12} + \frac{1}{8} + \frac{1}{6}\right)$

$= \frac{16}{30} \cdot \frac{10}{24} = \frac{2}{9}$ $\qquad tP_4 = \frac{8}{9}$

When t = 5, $P_5 = \frac{32}{6.5}\left(\frac{1}{4.3.2.1.}\right)$

$= \frac{32}{30} \cdot \frac{1}{24} = \frac{4}{90}$ $\qquad tP_5 = \frac{2}{9}$

Hence $c = \Sigma t\,P = \frac{1}{15} + \frac{5}{9} + \frac{7}{6} + \frac{8}{9} + \frac{2}{9}$

$= 2.9$

INPUT-TO-OUTPUT CHECKING

In this method tests are made at the test point starting nearest the input in the faulty area.

If component 1 is faulty
the number of tests necessary is 1

If component 2 is faulty
the number of tests necessary is 2

If component 3 is faulty
the number of tests necessary is 3

If component (n–1) is faulty
the number of tests necessary is (n–1)

If component n is faulty
the number of tests necessary is (n–1)

No additional test is required when considering the n th component because it is effectively checked by the (n–1) th test. Since the output is known to be incorrect, if the (n–1) th component is shown to be good, the fault must be in the n th component.

The mean number of tests required will again be $\Sigma t.P$. Then P is the same for all components, therefore $c = P\Sigma t$ where

$$P = \frac{1}{n}$$

$$\Sigma t = 1 + 2 + 3 \ldots\ldots + (n-1) + (n-1)$$

$$= \frac{n-1}{2}\left[1 + (n-1)\right] + (n-1)$$

$$= \frac{n(n-1)}{2} + (n-1)$$

$$= \frac{1}{2}(n-1)(n+2)$$

Therefore $c = \frac{1}{2n}(n-1)(n+2)$

CHAPTER 6 THE HALF-SPLIT METHOD

INTRODUCTION

In Chapter 5 it was shown that when it is required to narrow down a faulty area using a sequence of tests, the half-split method is the best method to use. In this chapter we will enlarge on the method and show how it can be applied to practical circuits. First we will state the rule and prove the formula which was used in Chapter 5.

THE HALF-SPLIT RULE

The half-split rule states that to locate a fault in a group of units or components, one of which is known to contain a fault, each check must be made at a point such that an equal number of components is eliminated whatever the result of the check.

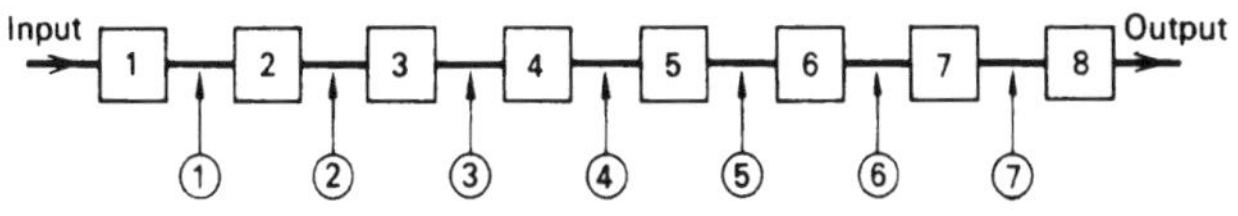

Figure 6.1 A simple series circuit

Consider the simple series circuit of Figure 6.1. If it is known to contain a fault, because the input is correct but the output is incorrect, the first check is made at test point 4. If an incorrect result is obtained the fault is in components 1, 2, 3, or 4, but if a correct result is obtained at test point 4 the fault is in components 5, 6, 7, or 8. Thus four components are eliminated whatever the result at test point 4. If the first check is made at any other point, an equal number of components is not eliminated with correct, and incorrect, results. A first check at test point 2, for instance, would eliminate two components if a correct result is obtained and 6 components if an incorrect result is obtained.

If the result of the test at point 4 is correct, the fault must lie in unit or component 5, 6, 7, or 8. The next test must therefore be made at the half-way point of these remaining components, i.e. at test point 6. If the result here is incorrect, for example, the fault must lie in 5 or 6, and one further test, at point 5, will determine which component is faulty.

In a simple application of the half-split rule, several assumptions are made. These are:

a. Only one component is faulty.

b. All components are equally reliable.

c. It is possible to check the signal at any point between components.

d. The test signal tolerances are known.

e. All checks are of the same difficulty and take the same time to complete.

The application of the rule and the validity and effect of the assumptions will be discussed later in this chapter. First we will prove the rule and determine the mean number of checks required to locate a fault when the rule is used.

PROOF OF THE HALF-SPLIT RULE

This proof involves simple differentiation, if the reader does not wish to follow it the result can be assumed.

The proof is given in the Appendix to this chapter, where it is shown that the number of checks needed to locate a fault, using the half-split rule, is 3.32 log n, where n is the number of components, and log is to base 10.

This is the formula which was used in Chapter 5 to compare the half-split method of fault location with other methods.

Let us put some values for n in the formula. When n = 2, c = 1, which is to be expected since only one check, between the two components, is necessary. When n = 8, c = 3; so that the faulty component can always be located by making only 3 checks. When n = 1,000, c = 10 (approximately); this compares with c = 500 (approximately) for the same n but using the beginning-to-end method. This last result shows how superior the half-split method becomes when large numbers of units or components are involved.

APPLYING THE RULE TO SERIES CIRCUITS

Referring again to Figure 6.1, let us assume that the fault is in component 2. The first check, as always, is at test point 4. This gives an incorrect result, indicating that the fault lies in components 1, 2, 3, or 4; and an equal number of components (5, 6, 7, and 8) are eliminated. The next check is made at test point 2, again halving the faulty area. This check again gives an incorrect result indicating a fault in component 1 or 2, and components 3 and 4 are eliminated. A final check at test point 1 gives a correct result so that the fault must be in component 2. Thus the fault is isolated using a sequence of three checks. A similar sequence is used if any one of the other components is faulty.

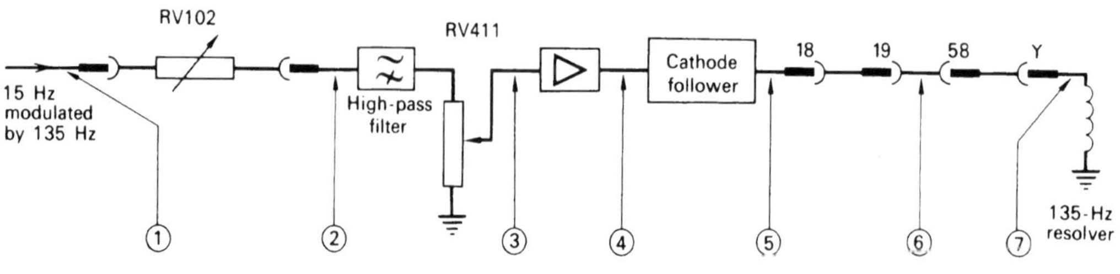

Figure 6.2 A practical series circuit

Now consider the practical example illustrated in Figure 6.2, which is an extract from a radar equipment diagram. If the 135-Hz resolver does not function correctly but the signal at test point 1 is correct, then the next check should be made at test point 4. We are assuming here that the test points divide the circuit into parts which have an equal probability of failure, and this assumption must be made if no contrary information concerning reliability is provided. If the result of the check at test point 4 is a correct signal, the next check must be made at either test point 5 or test point 6. It is not possible in this case to make an exact division into two, so the nearest possible division to the optimum is made. If the check at test point 5 indicates an incorrect signal then the fault lies in the cathode follower.

ASSUMPTIONS AND LIMITATIONS

In Section 6.2 we listed a number of assumptions which are made when the half-split rule is used. We will now investigate these in some detail to see whether they seriously restrict the application of the rule.

First, let us consider the assumption that only one component is faulty. This is not an unreasonable assumption because, when one component fails, an equipment should be taken out of operation for maintenance. It would not be usual to wait for two components to fail. Nevertheless if we consider conditions such as those that exist when an equipment has been in store, has been subjected to excessive environmental strain or has not been used since its construction, then it is possible for multiple faults to occur. However, when we apply the half-split rule to a series circuit with more than one fault, little modification is required. In these circumstances, application of the half-split rule will narrow down the faulty area to one unit or component. When the fault is corrected, it will be found that the circuit still does not function correctly. A second series of checks, starting with the same check as the first series, will then reveal the second faulty component, and so on until all faulty components are isolated. This can be easily seen by the reader if he assumes that multiple faults occur in Figure 6.1, and the half-split rule is applied to deduce the faulty components. It will be seen later that multiple faults cause more difficulty when they exist in circuits in which the components are not in series.

The second assumption was that all components are equally reliable. This means that when the components are split into two equal groups by quantity the probability of the fault being in one group is the same as the probability of the fault being in the other group. Again this is not an unreasonable assumption. Firstly, in the absence of contrary information, it has to be assumed that components have reliabilities of the same order. If this were not the case, fault finding by the reliability method might be a good alternative to the half-split method. Secondly, it is unlikely that the technician will know much about component reliabilities, other than large differences which he may have learned by experience; for example, he would suspect valves rather than resistors.

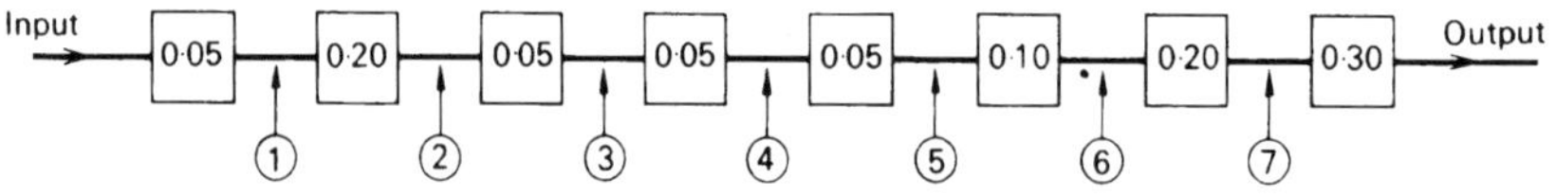

Figure 6.3 Series circuit showing failure rates

Nevertheless, it is possible to take reliability into account by modifying the basic rule such that the half-split divides the components into two groups of equal reliability. Consider the circuit of Figure 6.3. Each component has its reliability, measured as failure rate, in % per 1,000 hours, shown in the square. For convenience the sum of the failure rates has been made equal to one, which means that in the complete circuit 1% of the components will fail in a 1,000 hour period. Ignoring reliability we would choose test point 4 for the first check, but, taking reliability into account the first check must be at test point 6, such that the failure rate on each side totals 0.5. If this check gives an incorrect result, the next check must be at test point 2, which again is such that the failure rate on each side is equal.

For a practical example let us assume that the reliabilities of the parts illustrated in Figure 6.2 are as shown in Table 6.1. To halve the failure rate the first check must be made at test point 3. If this gives a correct result the second check must be made at test point 4, which, although it does not halve the reliabilities, is the nearest possible division.

At first sight test point 3 might be chosen for the first check, and 10 minutes would be required to reduce the faulty area to three units. The faulty area can however be reduced to two units by two checks, at test points 4 and 2, which take only seven minutes. In this case it is better to make two checks than one.

To summarize the assumptions which were made in the initial description of the half-split rule, we can say that the basic rule must, when the necessary information is available, be modified to take account of reliability and nature of the required checks. In practice the technician will only be required to deal with large differences in these factors. Fine distinctions are more appropriate to the preparation of aids to the task in the form of guides or automatic equipment.

RULES FOR MORE COMPLEX CIRCUITS

In all examples so far we have used a simple series circuit, not because the half-split rule applies only to this type of circuit, but because it is the easiest to start with. We have also seen that the biggest gain from using the half-split method is where there are a large number of units or components. Very few modern equipments have large numbers of components in series, and it is therefore necessary to supplement the basic rule with rules applicable to other common circuits. We will now discuss these additional rules.

Table 6.1: Failure Rates of Components in Figure 6.2

Components	RV102	Filter & RV411	Amplifier	Cathode Follower	Tag 18 Tag 19	Tag 58 Tag Y
Failure Rate (% per 1,000 h)	0.15	0.35	0.20	0.20	0.05	0.05

The third assumption which is worth some discussion is that all checks are equally difficult and take the same time to carry out. This assumption is important because the criterion of efficiency is generally taken as the time to locate a fault rather than the number of checks taken. On this basis a technician locating a fault using 10 checks in 30 minutes does better than one who uses 2 checks but takes 45 minutes. In practice the nature of a check must therefore be taken into account. As was the case with reliability, it is not possible to make fine distinctions between the difficulty of checks, but large differences can be dealt with as shown in the following example, which refers to Figure 6.4.

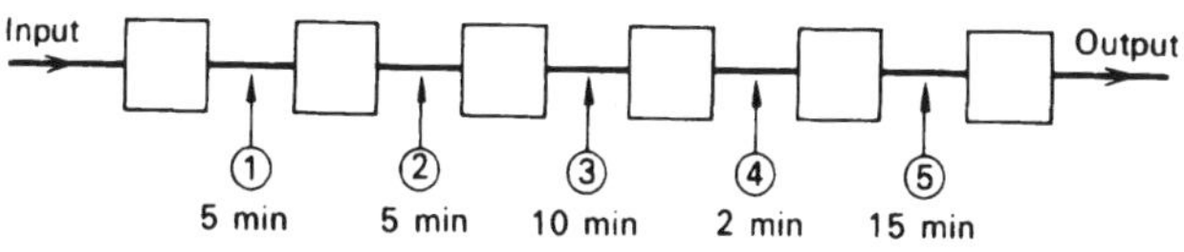

Figure 6.4 Series circuit showing checking times

DIVERGENCE

Divergence occurs when one input is the source of two or more outputs. It is probably the most common configuration because many equipments derive a number of outputs from a single input, such as a received signal or a power supply. The latter is probably the commonest example when it is remembered that every unit of an equipment usually derives its supply from one main source of power.

Divergence can occur at a point, such as a terminal post, or at a stage, such as outputs taken from the anode and the cathode of a valve. These two types are illustrated in Figure 6.5. Practical examples are illustrated in D3 and D4 of Figure 5.2.

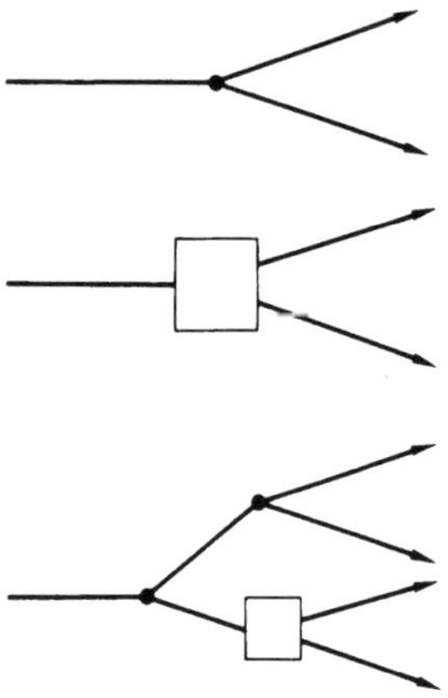

Figure 6.5 Types of divergence

For a divergent arrangement the rule is to check each output in turn, and to continue to search in the area common to the incorrect outputs. In the circuit of Figure 6.6, which illustrates part of a radar equipment block diagram, if there are incorrect signals at all speed, miles flown and drift outputs, but there are correct signals at the approximate speed neons and memory lamp, in what area does the fault lie? Clearly the receiver chain must be functioning correctly to give the two correct outputs, and since the fault must be in an area common to the incorrect outputs, it must lie between the output of the receiver chain and some point in the speed chain which is common to all incorrect outputs.

In Figure 5.2 the two main outputs are the audio signal to the headphones and the test meter reading. Applying the rule for divergence, if both are incorrect the fault lies in, or before, the second i.f. stage. If the meter is correct, but there is no audio signal, the fault lies between the second i.f. output and the headphones.

It is worth noting that it may not be necessary to check all available outputs. In the Figure 6.6 example it is sufficient to know that the speed and drift meters are incorrect to obtain the same result. Of course, if the other outputs are readily available, there is no harm in using them to confirm deductions. The reader can use Figure 6.6 to practise the application of the rule by assuming various combinations of correct and incorrect outputs. A point to remember in doing this is that where several separate faulty areas occur there must be more than one fault; for instance, if the approximate speed neons are correct, and the memory lamp is incorrect, a fault lies between the signal noise chain output and the memory lamp. If at the same time the speed meter is correct, but the drift meter and drift to GPI are incorrect, another faulty area must be between the speed chain output and the azimuth chain output. The two faulty areas are separate so there must be two separate faults.

CONVERGENCE

Convergence occurs when one output is derived from two or more inputs; this is illustrated in Figure 6.7. Convergence is not as simple to deal with as divergence because the faulty area cannot be deduced from the single output. Before deciding on how checking is to be done we must know how the inputs are combined. There are two possibilities. First, the output may be produced only when all the inputs are present; this is called **summative convergence** and is equivalent to the AND gate in computer terminology. Second, the output may be produced when only one of the inputs is present; this called **alternative convergence** and is equivalent to the OR gate in computer terminology. The two types of convergence are treated in a slightly different manner so that the first point about this type of circuit, and indeed about all circuits, is that the technician must know the function of the circuit. However this does not mean that he must have a deep knowledge of how the circuit works in terms of electronic theory.

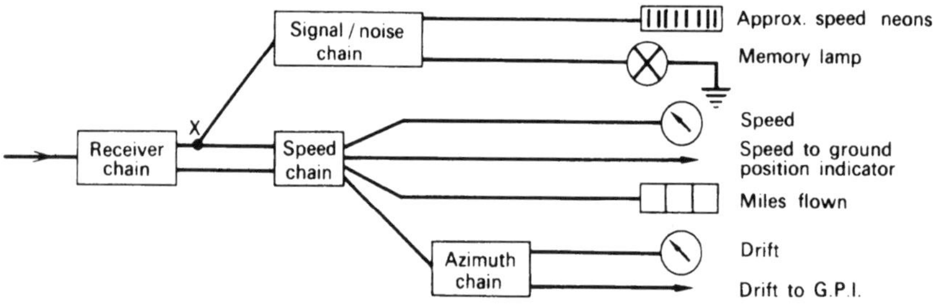

Figure 6.6 A practical divergent block diagram

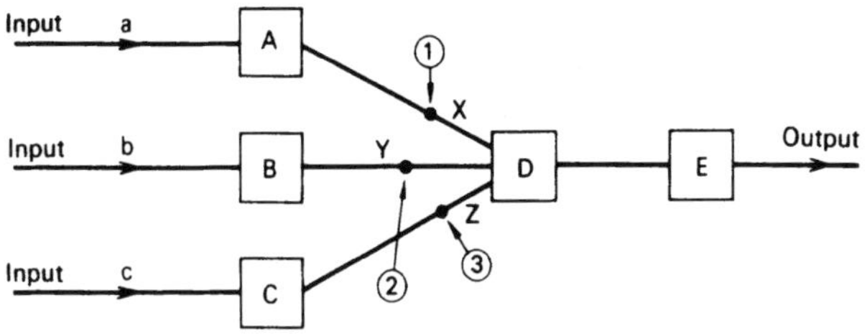

Figure 6.7 A convergent circuit

There are two examples of summative convergence in Figure 5.2. In square B3 the r.f. signal and the local oscillator signal are combined to form the i.f. output. In square D3 the b.f.o. and the i.f. signals are combined to produce the audio output. In each case, both signal inputs must be present to produce a correct output.

The rule for summative convergence is to check the inputs near to the point of convergence, in turn. If any one is incorrect, the fault lies in that input; if all are correct, search at, or after, the point of convergence. Referring to Figure 6.7, if a fault exists in this circuit the first check is made at test point 1. If this gives an incorrect result the fault lies between the input 'a' and test point 1; if it gives a correct result, check at test point 2; and so on. If the checks at test points 1, 2 and 3 all give correct signals the fault lies in, or after, circuit D.

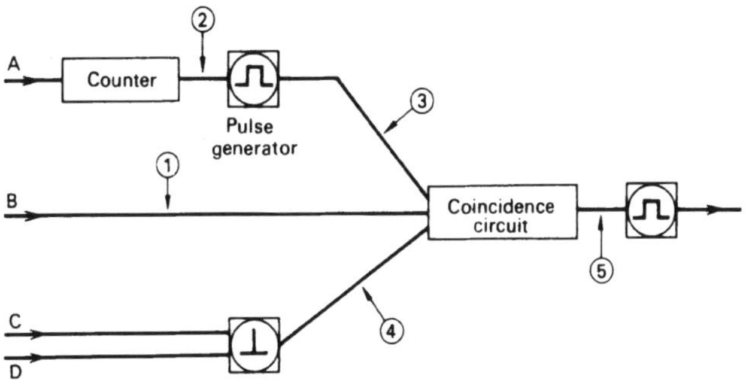

Figure 6.8 A practical convergent circuit

A practical example is shown in Figure 6.8. The first check must be made at test point 3 and, depending on the result of this check, subsequent checks will be made in either the incorrect input circuit or in the coincidence and succeeding circuits.

Alternative convergence rarely occurs except in computer circuits, since switches are usually used when one of a selection of inputs is required. The rule for switching circuits is dealt with later in this Chapter. The diagram for a circuit containing alternative convergence is the same as that for summative convergence, but the function of the circuit is different. Figure 6.7 can therefore also represent an example of alternative convergence. The rule is to check the output with each input present in turn. If more than one input gives incorrect outputs the fault lies after the point of convergence; if only one input gives an incorrect output the fault lies before the point of convergence.

Applying this rule to the circuit of Figure 6.7, the first check is with input 'a' present; whatever the result of this check the next check is with input 'b' present. If the results of both checks are correct, make the next check using input 'c'. If one result is incorrect, check between the input which was used for that check and the point of convergence. If both results are incorrect check between the convergent point and the output. Thus if there is a fault in unit D, the sequence would be: check the output using input 'a', which will give an incorrect result; check the output using input 'b' again giving an incorrect result. If there is a fault in unit C, the sequence would be: check the output using input 'a', which will give a correct result; check the output using input 'b', again giving a correct result; then check the output using input 'c' which will give an incorrect result.

FEEDBACK

Feedback exists where a signal is connected to a point which is earlier in its flow path. This type of circuit makes fault finding difficult because the signal flow is in a closed loop. If a fault exists in the loop the signal is incorrect at all points in the whole loop and no narrowing down of the faulty area can be achieved simply by making additional checks. There is no starting point and no finishing point. The feedback circuit is exemplified in Figure 5.2 by the a.g.c. path (D5, A3, B3, and C3). The complete loop contains the two i.f. stages, the a.g.c. circuit, the screened cable, the r.f. stage and the mixer. If an incorrect signal is obtained at any point in this loop, there must be an incorrect signal at all points in the loop and nothing can be gained from additional checks within the loop. It is therefore necessary to change the feedback circuit in some way if we are to locate faults in it. The rule for feedback circuits is therefore: change the feedback circuit.

The nature of the change will depend on the function of the circuit, so once again the technician must understand the function of the circuit. The first question to ask is: what type of feedback circuit is it? There are two types: **modifying** feedback and **sustaining** feedback.

Modifying feedback is such that a signal is fed back with the object only of modifying the output. For example, the modification could be a change in gain as in an a.g.c. feedback circuit, or a change in frequency as in an a.f.c. circuit, or a change in wave shape. In all cases the output would be present without the feedback but it would, of course, not have the same characteristics as the output with feedback. In applying the rule to a feedback circuit of the modifying type, the change could be complete disconnection, earthing the feedback path or operating a switch.

With **sustaining feedback** the feedback path is essential for the production of an output, and simply disconnecting the loop would result in no output. The type of change which has to be made is therefore disconnection of the loop and injection of a signal to represent the one which has been disconnected. The most common example of sustaining feedback is the simple oscillator, which relies on positive feedback for its operation. Servo-mechanisms are in another class of circuit which may often include sustaining feedback.

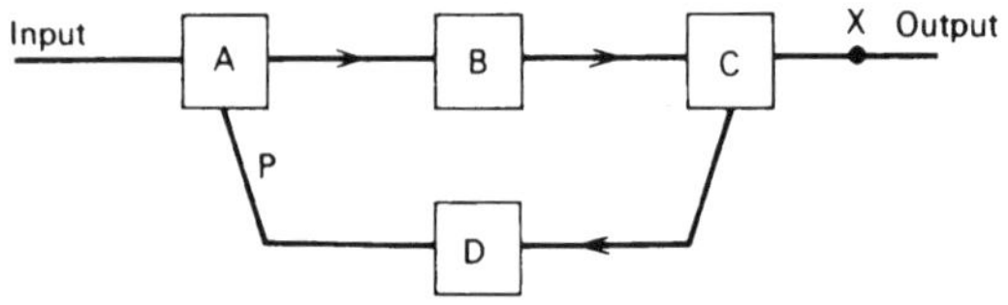

Figure 6.9 A feedback circuit

For an example of the application of the feedback rule consider Figure 6.9. If the input is correct and the output is incorrect, the feedback loop must be changed in order to narrow down the faulty area. Let us assume that the feedback is of the modifying type so that we can disconnect the loop at point P. This is the point at which the feedback signal is connected to the main path, and this is always the best place for a disconnection. The circuit now becomes that shown in Figure 6.10, which can be re-drawn in the series form shown at Figure 6.11. The output is again tested, remembering that its characteristics will have changed now that the feedback has been removed. A correct output indicates that a fault lies between a point in circuit C and the point P; an incorrect output indicates that a fault lies between the input and the output, excluding the feedback loop.

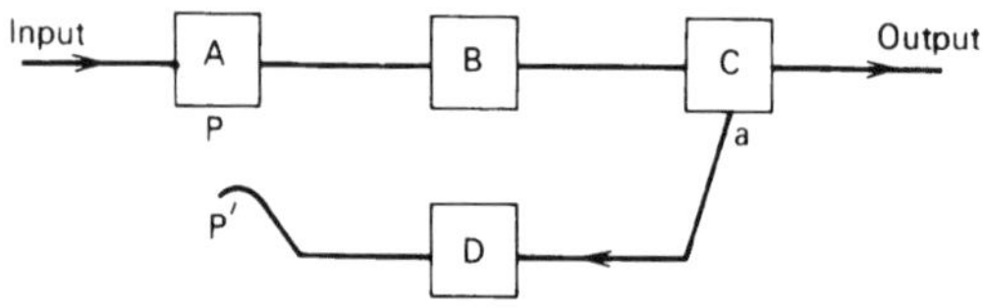

Figure 6.10 A modified feedback circuit

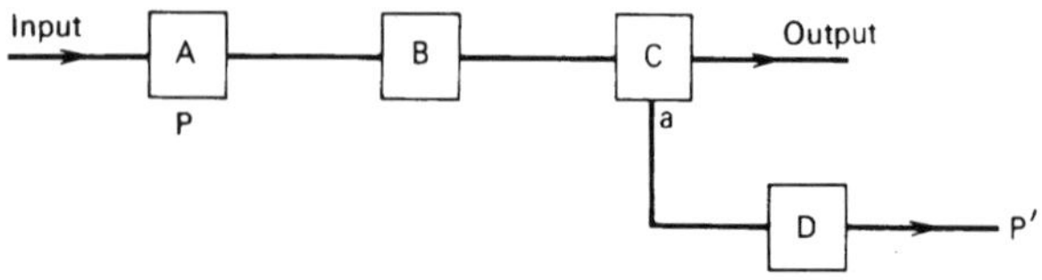

Figure 6.11 Series form of Figure 6.10

Had the above example been for a sustaining type of feedback loop, it would have been necessary to disconnect the feedback loop at P and then inject a signal into stage A corresponding to the normal feedback signal. A correct output would indicate a fault in the feedback loop and an incorrect output would indicate a fault between the input and output, but excluding the feedback loop.

The applications of feedback are many and varied so that the above examples give only a simple picture of the treatment of feedback loops. For the technician the important points about feedback are being able to recognize it, understanding the function of the circuit, and knowing how to apply the basic rule.

SWITCHING CIRCUITS

Switches are commonly used to modify the function of an equipment and, as such, they provide an excellent means of narrowing down a faulty area. They are usually easily accessible, simple in operation and result in a clearly defined change in output characteristics. There are very many types of switches, varying in mechanical operation and electrical connection. The important point in fault location involving a switch is, once again, to know the function of the switch, that is to know how its operation changes the structure of the equipment.

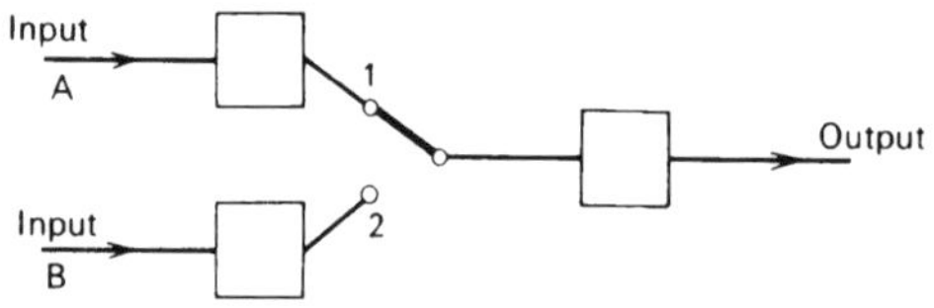

Figure 6.12 A switching circuit

The rule for switching circuits is: change the switch position. If the fault symptoms remain, the fault is in the area common to the two switch positions. If the fault symptoms go, the fault is in the area which has been switched out. In the theoretical example of Figure 6.12, if the two inputs are correct, but the output is incorrect, the first check is to change the switch position. If the output is still incorrect the fault must lie in the region between the pole of the switch and the output. If the output is correct, the fault must lie between input A and position 1 of the switch.

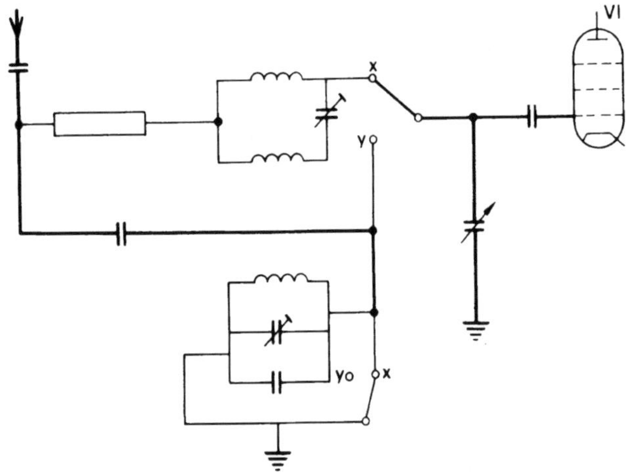

Figure 6.13 A practical switching circuit

Figure 6.13 shows a practical example. If there is no input to V1 on wave band x, we would switch to wave band y. If there is still no input to V1 the fault must lie in the areas common to both switch positions i.e. in the part of the circuit indicated by the heavy lines in Figure 6.13. If there is an input to V1 when the switch position is changed from x to y the fault must lie in the area which is only used when wave band x is selected.

APPLYING THE RULES TO COMBINED CIRCUITS

The rules discussed above have been exemplified using small isolated circuits. In practice an equipment is made of a large number of components and units connected in various combinations of the simple structures. Fault location will probably also involve carrying out a series of pre-determined front panel checks, which are made by the technician before he starts his location method. Front panel checks are discussed in Chapter 9.

between the inputs and the pole of the switch i.e. in the area common to the incorrect outputs. Next, we must remember that for all checks the basic aim is to halve the area containing the fault, so, applying the half-split rule, the area can be divided into two areas containing the same number of units by making a check at point A. In the absence of any reliability data we assume that all units have an equal failure rate. If this gives a correct result, the fault lies between point A and the pole of the switch. Here we have a feedback circuit and, depending on its type and function, it is necessary to change it before making any more checks. If the feedback is of the sustaining type, we now disconnect at point A and inject a suitable signal into the main flow path. If the outputs are now all correct, the fault is in unit B. If the outputs are incorrect we have a series chain of three units (C, D and E) to contend with, and, since there are only two possible test points, this leaves no choice of optimum test point.

When we have discussed functional checks in Chapter 9, we will show how the rules can be applied to the equipment illustrated in Figure 5.2. At this stage it is sufficient to study the diagram and to note that it consists of the circuit types which have been mentioned in this Chapter.

DIFFICULT FAULTS

It is not surprising, when the nature of electronic equipment is considered, that there are exceptions to the rules described in this chapter. The fact that exceptions exist means that when the rules are apparently failing the technician must look for the sort of conditions mentioned below; fortunately, in a well designed equipment, they should be rare. The main exceptions occur in the following circumstances.

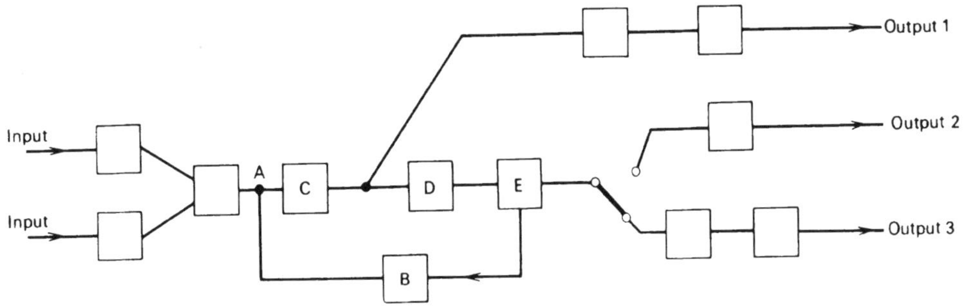

Figure 6.14 Equipment block diagram

The block diagram of a practical equipment may be similar to that illustrated in Figure 6.14. Suppose that the initial checks, which include the operation of the switch, show all outputs to be incorrect, and it is known that the two inputs are correct. These symptoms tell us that the fault lies

a. **Multiple faults.** The rules are based on the assumption that only one fault exists at a time. There will therefore be apparent inconsistencies in the results of checks when more than one fault exists. For example, referring to Figure 6.6, if all outputs are incorrect, we would deduce that the fault is in the part of the circuit preceding the earliest divergent point (X), but the same symptoms would occur if there were faults in all arms of the divergent parts of the circuit. Application of the rule will thus lead to an incorrect deduction. In Section 6.5 it was seen that the multiple faults do not affect the procedure where series circuits are concerned. For more complex circuits, once the fact that multiple faults exist is recognized, the difficulty of fault finding is not unduly increased. In the above example using Figure 6.6, we could assume that two faults are present and so look for two separate areas which give the incorrect outputs. The two possible areas in the example are between point X and the signal/noise chain input, and between the receiver chain output and the speed chain input. Each suspect area can then be dealt with separately.

b. **The fault connects otherwise functionally separate paths.** This occurs mainly where the fault is a short circuit, for example between leads in a cable or between the pins of a plug. The fault alters the structure of the equipment by adding an unwanted connection and incorrect deductions will be made because the circuit no longer corresponds with the diagram. As with multiple faults, this type of fault is soon recognized by the failure of the normal rules. It can best be overcome by changing to another systematic method of elimination.

c. **Fault symptoms are reflected back to an earlier part of the circuit.** This situation usually occurs when the fault is a short circuit and when the circuits are d.c. coupled. If the fault in one arm of a divergent circuit with two outputs is reflected further back than the divergent point both outputs will be incorrect. This leads to an incorrect deduction. This type of fault can be avoided by a design which only provides test points and other maintenance facilities at points where circuits can be isolated. As with the type of fault discussed in (b) above, the condition is easily recognized and can be overcome by changing to another systematic method of elimination.

d. **The fault is catastrophic.** This is an extension of (c) to the stage where the equipment becomes completely inoperative because a component fails, for example when the fuses always blow on switching on. Once again a change of method is required, but there is no necessity to resort to inefficient methods. In the fuse example, disconnection of half the circuit, replacing the fuse and switching on again would be a logical method, as would be a systematic visual inspection of the circuit.

e. **The circuit is such that parts of it are influenced by different parts of the signal.** For example, a certain trigger circuit may be responsive only to the amplitude of an applied pulse, whereas another trigger circuit in the equipment may respond only to the duration of the same pulse. This situation can cause difficulty in fault location. The difficulty lies more in correctly specifying a signal's characteristics than in the application of the fault location rules. The problem of signal specifications and tolerances is discussed in Chapter 10.

We have seen that, where the rules for fault location cannot be applied in the form stated, the technician may still use one or more of the other methods described in Chapter 5. It is still possible to be systematic and to use a check which halves the faulty area.

SUMMARY

If an equipment is constructed such that the technician must use a sequential method of fault location, the best method is the half-split. The half-split rule is:

> Make every check at a point such that an equal number of components is eliminated whatever the result of the check.

The optimum place to check, as determined by the rule, can be modified to take account of:

a. The reliability of the components.

b. The ease with which a check can be made.

Supplementary rules to take account of more complex circuits are:

a. **Divergence.** Check each output in turn and continue to search in the area common to the incorrect outputs.

b. **Summative Convergence.** Check the inputs near to the point of convergence in turn. If any one is incorrect the fault lies in that input; if all are correct, search at, or after, the point of convergence.

c. **Alternative Convergence.** Check the output with each input present in turn. If more than one input gives incorrect outputs the fault lies after the point of convergence. If only one input gives an incorrect output the fault lies before the point of convergence.

d. **Feedback.** Change the feedback circuit.

e. **Switching Circuits.** Change the switch position. If the fault symptoms remain, the fault is in the area common to the switch positions. If the fault symptoms go then the fault is in the area switched out.

APPENDIX TO CHAPTER 6

PROOF OF THE HALF-SPLIT RULE

For this proof the series circuit of **n** components or units, shown in Figure 6.15, is used, and all the assumptions listed in Section 6.1 are made.

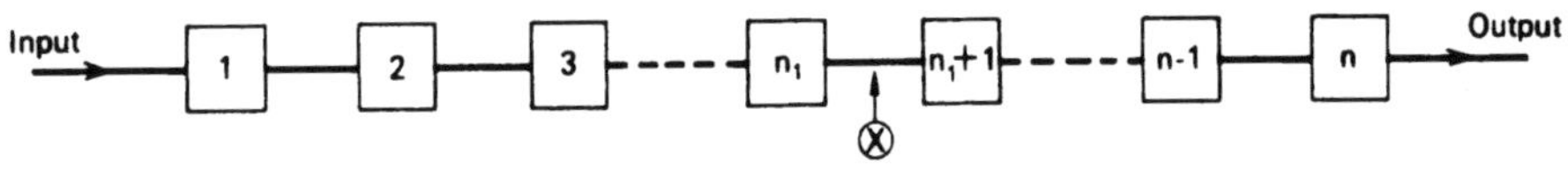

Figure 6.15 A simple series circuit

We will start by making a check at an arbitrary test point x, which divides the circuit into n_1 and $(n - n_1)$ components. If the result obtained at x is correct, the fault must be in a group of $(n - n_1)$ components; if the result obtained at x is incorrect the fault is in a group of n_1 components. We want to know where x should be, in terms of n_1, so that the fault is located in a minimum mean number of checks. ı ne reasons for using a **mean** number of checks are given in Chapter 5. We also assume that, having made the initial selection at test point x, some predetermined rule is used to complete the sequence of checks; this rule need not be specified but, to simplify the proof given here, we will assume that the rule for input-to-output checking is used. This rule is explained in Chapter 5. For n components in series it gives a mean number of checks, c, of $\frac{1}{2}n\,(n-1)\,(n+2)$. Thus the number of checks which are required to locate the fault in a group of n_1 components is $\frac{1}{2}n_1\,(n_1 - 1)\,(n_1 + 2)$, and the number of checks which are required to locate the fault in a group of $(n - n_1)$ components is

$$\frac{1}{2(n-n_1)} \cdot (n-n_1-1)\,(n-n_1+2)$$

In Figure 6.15, the probability that the fault is in the set of n_1 components is

$$\frac{n_1}{n}$$

and the probability that the fault is in the set of $(n - n_1)$ components is

$$\frac{n-n_1}{n}$$

If we add the number of checks times the probability of that number being required, for the two parts of the series, then the overall mean number of checks required to locate the fault is obtained. We must also remember to add the first check which is made at point x. Thus, for a series of n components when a first check is made at point x and subsequent checks are made according to the input-to-output rule, the mean number of checks required is:

$$c = 1 + \frac{n_1}{n} \cdot \frac{1}{2n_1} \cdot (n_1-1)\,(n_1+2) + \frac{(n-n_1)}{n} \cdot \frac{1}{2(n-n_1)} \cdot (n-n_1-1)\,(n-n_1+2)$$

$$c = 1 + \frac{1}{2n} \cdot (n_1-1)\,(n_1+2) + \frac{1}{2n} \cdot (n-n_1-1)\,(n-n_1+2)$$

We can differentiate this expression with respect to n_1 and put the result equal to zero to obtain the value of n_1 which gives a minimum value for c.

Differentiating with respect to n_1, $\frac{dc}{dn_1}$ becomes

$$\frac{1}{2n}\,(n_1-1) + (n_1+2) - (n-n_1-1)\,(n-n_1-2)$$

$$\frac{dc}{dn_1} = \frac{1}{2n}\,(4n_1 - 2n) = \frac{2n_1 - n}{n}$$

Putting $\frac{dc}{dn_1}$ equal to zero,

$$0 = \frac{2n_1 - n}{n}$$

which gives a practical solution of $2n_1 - n = 0$,

or $n_1 = \frac{n}{2}$

The condition for a minimum mean number of checks is therefore that the initial check is made in such a position that it divides the components in half. In other words, whatever the result of the check may be, half of the components are eliminated.

There is no difference in concept between the first check and any subsequent check, so that the rule governing the optimum division of components applies at all stages. Thus, for every check, the checking position is chosen to be half way between the remaining components.

We can now calculate the number of checks which are required to locate a fault using the half-split rule. If c is the number of checks which are required to locate one faulty component in a group of n components, then because each check divides the number of components successively by two, we must have $2^c = n$. Taking logarithms to base 10:

$$c \log 2 = \log n$$

$$c = \frac{\log n}{\log 2}$$

Hence $c = 3.32 \log n$.

CHAPTER 7 AIDS TO MAINTENANCE

INTRODUCTION

In discussing factors affecting maintenance in Chapter 4, the impact of aids on the requirements of the maintenance task was emphasized. It was pointed out that aids can transform the technician's work from being extremely difficult when no aids are provided, to very mundane when advanced automatic test equipment is available. In this chapter we shall investigate more fully the reasons for providing aids, the nature of the information provided by aids, and how the information is presented.

We will define an aid as something which is incorporated in, or provided with, an equipment, but which is required only for maintenance purposes. Conversely we can say that aids do not form part of the operational requirement as defined in Chapter 4. Clearly items such as test meters and oscilloscopes satisfy this definition and are aids; test points and indicator lamps which are part of an equipment, but which are provided for maintenance only and not for operation, must also be classified as aids. Indicator lamps, meters, plugs and sockets which are required primarily for the operation of the equipment cannot be classified as aids to maintenance, though they may well be used during a maintenance task.

An important point about aids, which is often overlooked, is that the information provided by them should be strictly related to that required for carrying out the work. It is often the case that a great deal of information is provided irrespective of its relevance and this is particularly true of written aids. If the technician has to search amongst a mass of irrelevant data for the information he requires, then his job is unnecessarily made more difficult than if he had to search amongst a much smaller amount of data all related to his work. There is no need, for instance, for the voltage at every point in a unit to be specified when one or two voltages would permit fault location to a single component. It is also true that, on occasions, the information required for a maintenance task is not provided. The technician's task in these circumstances is again made more difficult because he has either to search elsewhere for information, or deduce it himself if his training fits him to do so.

WHY AIDS ARE NECESSARY

If an equipment were perfect, it would have an infinite m.t.b.f., there would be no need for maintenance and no need for aids. We know this is not possible, so it is necessary to make provision for maintenance by supplying **facts** about the equipment which are necessary for both routine and non-routine tasks, and by giving **guidance** on maintenance procedures, i.e. by specifying what facts are necessary for particular purposes. As an example, the **facts** provided about an i.f. amplifier could be: the supply voltage; the voltages at valve pins; and waveform specifications for the input to the amplifier and the output from each stage. **Guidance** could be a set procedure for checking the amplifier; for example: check the input signal; check the supply voltage; check the signal at the output from V4.

Facts and guidance could be given during training but the memory load would be so great that few people would be able to assimilate them. The technician must therefore be given facts and guidance in the form of aids.

THE PROVISION OF FACTS

The facts about electronic equipment which are required for maintenance come under the headings: facts concerning the electrical structure, which are information about the components and how they are connected; facts concerning the mechanical structure, which are information about how the components are laid out and mounted; facts concerning dynamic measurements, which are the voltages and waveforms expected under normal operating conditions; and facts concerning static measurements, which are measurements taken when the equipment is not working. Examples of factual information about a transistor oscillator are given in Figures 7.1, 7.2 and 7.3. Figure 7.1 illustrates the electrical structure; it shows that resistors, capacitors and a transistor are used, and how they are connected. Figure 7.2 shows that the components are mounted on a printed circuit board and also shows the position of each component. Figure 7.3 illustrates the waveform that should appear at the output under normal operating conditions. The important static measurements are included in Figure 7.1 as component values.

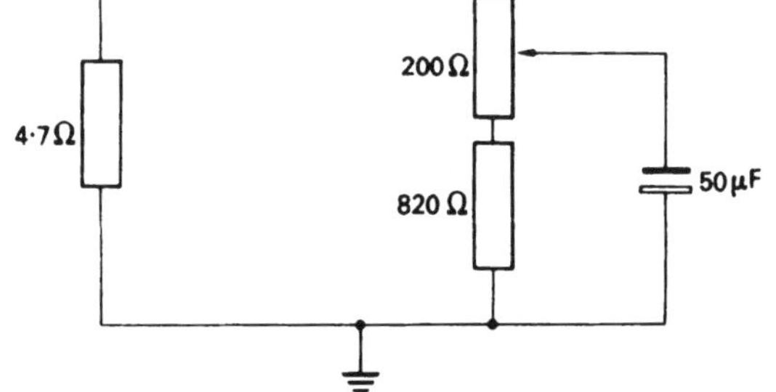

Figure 7.1 Transistor oscillator: electrical structure

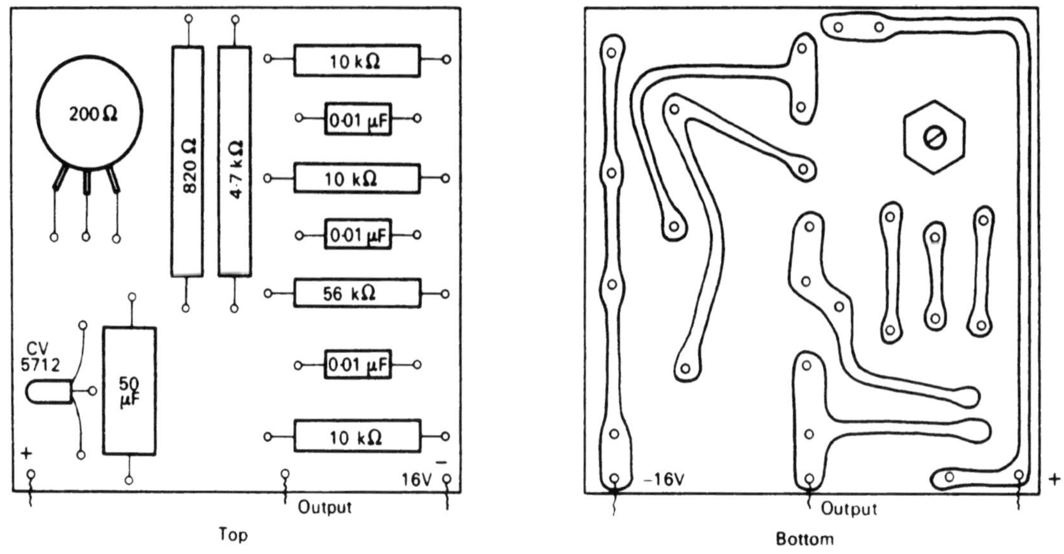

Figure 7.2 Transistor oscillator: mechanical structure

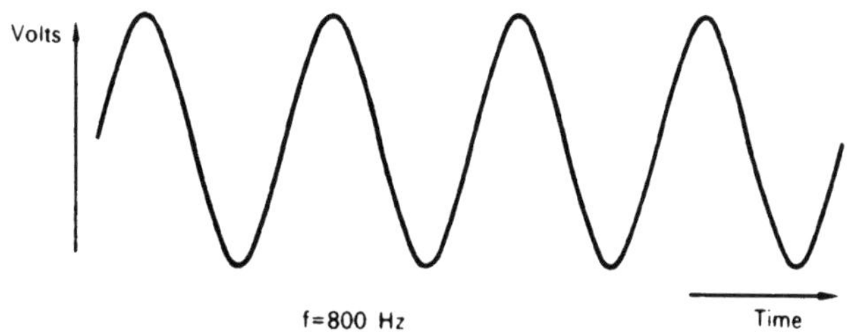

Figure 7.3 Transistor oscillator: waveform

Perhaps the most difficult information to obtain is that concerning dynamic measurements. Not only does the equipment have to be constructed in such a manner that information about performance can be obtained, but test equipment must also be provided which can reveal the information. In other words, an accessible test point must be provided together with a device for making a measurement at the test point.

Another difficulty arises in specifying tolerances. For example, if a supply is specified as 12V, how much deviation is permissible before operational performance is degraded? Test equipment and the problem of tolerance are more fully discussed in Chapter 10.

THE PROVISION OF GUIDANCE

Technicians are not expected to work out for themselves when checks on an equipment are to be made and what checks are to be made; so that guidance is usually provided. For preventive maintenance, schedules are always available. An example of one of these is given in Table 7.1. It consists of a simple series of checks which are to be carried out in order, and which require no decision concerning the next step. For corrective maintenance, guidance is required for fault detection and fault location. However, the task is not always to carry out the same pre-determined series of checks, so the guidance given is more complicated in form than that given for preventive maintenance. The actual form will depend on the method of fault location which is most appropriate for the equipment, and may also depend on the fault symptom. The method may be the half-split, or any of the other methods described in Chapter 5, and it will determine where to check; factual information should also be available on what to check at each point. The provision of instructions for fault detection and location is discussed fully in Chapter 9 under the heading 'Fault Finding Guides'.

Table 7.1: Preventive Maintenance Schedule

100h, 200h and 600h servicing schedules for airborne radar Type XY

ITEM NUMBER	ITEM	100h	200h	600h	OPERATION
1	Scanner motors (azimuth drive, roll, pitch and tilt)	*	*		Remove for overhaul and substitute serviceable motors.
2	Scanner system	*	*		Examine for corrosion, damage and security. Remove heading contact cover plate. Remove and clean heading contacts. Replace contacts and cover plate.
				*	Remove for overhaul and substitute serviceable system.
3	Waveguide, flexible coupling, support brackets	*	*	*	Examine for corrosion, damage and security.
4	Indicating unit	*	*		Remove blower motor for overhaul, substitute serviceable motor.
5	Junction box	*	*	*	Examine for corrosion, damage and security. Clean relay contacts and examine for pitting.
6	All connectors				Examine colour coding for deterioration.
7	All plugs and sockets	*	*	*	Examine for corrosion, damage, security of attachment and correct locking. Coat male threads with grease.
8	Co-axial connectors			*	Check that insulation resistance is not less than 10 megohms.
9	Dessicators	*	*	*	Examine crystals for discolouration. Change if light blue or pink.
10	Functional check	*	*	*	Carry out full functional check as specified in Chapter 9.
11	Recording	*	*	*	Complete and sign appropriate servicing forms.

PRESENTATION OF INFORMATION

Having decided what facts and guidance must be provided for maintenance, the problem arises of how best to present the information. The commonest and simplest method is by means of written information in equipment handbooks. For our purposes, all information which can be presented in a handbook, including all forms of diagram, will be referred to as 'written'. Much less common is the fully automated presentation used in automatic test equipment. Between these extremes there are many variations and combinations of methods of presenting maintenance information.

The circuit diagram is the best known written aid, and it has the widest use. Its main purpose is to provide information about the electrical structure of an equipment, and it is supplied with almost all electronic equipment. The way in which a diagram is drawn depends upon the task for which it is required. We would not expect a diagram produced as an aid to manufacture, for instance, to look the same as one produced as an aid to fault location. The method of drawing diagrams in a form which is most suitable for fault location is discussed fully in Chapter 8.

Other forms of written aid are: tables of voltages; waveforms; structural drawings; and photographs.

We call aids which cannot be incorporated into a handbook **hardware** aids. The presentation of maintenance information in hardware aids may be visual, in which film or television is used; or, less commonly, aural, in which a tape recorder and earphones are used. Some hardware aids use both aural and visual presentation; an example is a commercially available device which comprises a desk mounted visual unit plus a tape deck and control unit. Instructions are given aurally via earphones and visually using film. The device has widespread application and is being used for routine servicing, fault location, repair and assembly. When a hardware aid does more than merely present facts at the request of the operator, it can be considered to be automatic test equipment. This is fully dealt with in Chapter 10.

The choice of the method of presentation of aids to maintenance is a maintenance policy decision, and it depends on the factors discussed in Chapter 4. An example is the choice of aural aids to maintenance of aircraft instrument systems. The confined working space of an aircraft cockpit makes the use of handbooks difficult, but information can be passed fairly easily to the technician in aural form via earphones. In addition, this method leaves both hands free for other tasks. Another example, which is not visual or aural, is provided by computers. Since they perform complex functions on input data according to pre-determined instructions, then the best aid to maintenance of a computer is provided by special input data and instructions which can test the functions of the computer in stages suitable for locating faults.

OTHER FORMS OF AID

It would not be possible to carry out the maintenance task without access to many forms of aid other than the test equipment and written aids mentioned above. A hand tool, for instance, is an aid to the maintenance task, as is assistance from superiors. Another form is the maintenance record of the equipment, which, if properly kept, can provide useful information on previous unserviceability. The importance of meticulous recording of maintenance tasks has already been stressed. The aids discussed fully in this book are new types of diagram, fault finding guides and test equipment.

SUMMARY

Aids to maintenance give facts about the equipment and guidance concerning maintenance procedures.

Facts can relate to the electrical structure of the equipment, to the mechanical structure, to dynamic measurements and to static measurements.

Guidance takes the forms of checking schedules for preventive maintenance and fault finding guides for corrective maintenance.

Aids can be presented in the form of handbooks, including diagrams; these are written aids. Circuit diagrams, tables of voltages and waveforms are written aids. Aids can also be visual or aural, culminating in automatic test equipment.

CHAPTER 8 FUNCTIONAL DIAGRAMS

INTRODUCTION

Diagrams are the main aids to maintenance, and it is therefore essential that they contain information which is relevant to the jobs for which they are provided, and that the information is presented in the best possible manner. For fault location the most essential information is that concerning functional structure i.e. how the components are connected to perform their required function. Diagrams which illustrate functional structure in a simple and useful manner are called **functional diagrams**. Figure 5.2 is an example of a functional diagram.

A brief inspection of Figure 5.2 tells us that the function of the equipment is to convert signals received by the aerial into an audio output, and that this is done in a series of amplifying, mixing and detecting stages with feedback at appropriate points. Chapter 5 gave a description of the circuit in functional terms, and in this chapter we will see how Figure 5.2 satisfies the definition and rules for the production of functional diagrams. It is of interest to note that a functional description does not require detailed theoretical explanations of circuit operation; it is a description of what the device does, not how it does it in theoretical terms.

This chapter reviews the development of functional diagrams, describes in detail how they should be drawn, with examples, and how their use simplifies fault location.

DEVELOPMENT OF FUNCTIONAL DIAGRAMS

In the past there have been a number of occasions on which circuit diagrams have been criticized on the grounds of bad presentation, and there have also been attempts to specify the requirements of a good diagram. Unfortunately most of these efforts have resulted in little or no change in drawing practice. It is only recently that it has been shown how a wide range of complex electronic equipment diagrams can be greatly improved by the application of certain fundamental principles.

Bainbridge-Bell in 1953 put forward the idea that the main path in a circuit is made up of a number of links in a chain joining an input cause to an output effect. He suggested that the operation of a circuit is most easily followed if the chain is as straight as possible, since the reader's eye dislikes sudden changes in direction when scanning the diagram. Since we read from left to right, the cause-to-effect path should go from left to right with any deviation clearly marked. Bainbridge-Bell also suggested other changes, such as giving less emphasis to following the mechanical structure of the circuit being illustrated, and adopting more liberal use of sloping lines.

The British Standards Institution has also indicated where the emphasis should lie. BS530:1948 (Graphical symbols for telecommunications) states: 'Diagrams should be drawn so that the main sequence of cause-to-effect goes from left to right, and/or from top to bottom. The input (e.g. aerial of receiver) should always be on the left, and the output (e.g. aerial of transmitter) on the right. When this is impracticable, the direction of operation should be shown by an arrow. Balanced circuits are permitted exceptions from these recommendations. Components associated with each operational stage should be grouped together'.

A more recent standard, BS3939:1966 (Graphical symbols for electrical power, telecommunications and electronics diagrams) does not contain guiding principles for drawing, but it does define a circuit diagram as: 'A diagram which depicts by means of symbols the components and their interconnections concerned in the operation of a circuit. The aim should be to show the operation of the circuit as clearly as possible and therefore circuit diagrams do not necessarily depict spatial relationships of the various items and their connections'. This indicates that emphasis should be on the clear illustration of function rather than mechanical structure.

Another professional body which has indicated a need for improved drawing standards is the Institution of Electronic and Radio Engineers. In a report of their Education and Training Committee published in 1964 it was stated that professional engineers do not pay sufficient attention to: 'Laying out a circuit diagram. The fundamental principle, which should be observed, but frequently is not, is that the layout should show the function of the various parts of the circuit and of the circuit overall. It is very useful when a circuit is laid out as a more detailed version of a block schematic. Clarity should not be sacrificed for the sake of neatness or appearance'.

These examples give some indication of the characteristics of good diagrams, but unfortunately, really good circuit diagrams are still rare.

DEVELOPMENT OF DIAGRAMS IN THE ROYAL AIR FORCE

In Technical Training Command of the Royal Air Force, interest developed during the preparation of material for experimental fault diagnosis training courses, which were run with the object of determining some of the critical factors which occur in teaching airmen to diagnose faults. In these courses a number of inexperienced airmen were taught to diagnose faults on two moderately complex airborne radar equipments. A specially developed course was used, and it resulted in these airmen reaching a satisfactory standard in a period considerably shorter than that normally required.

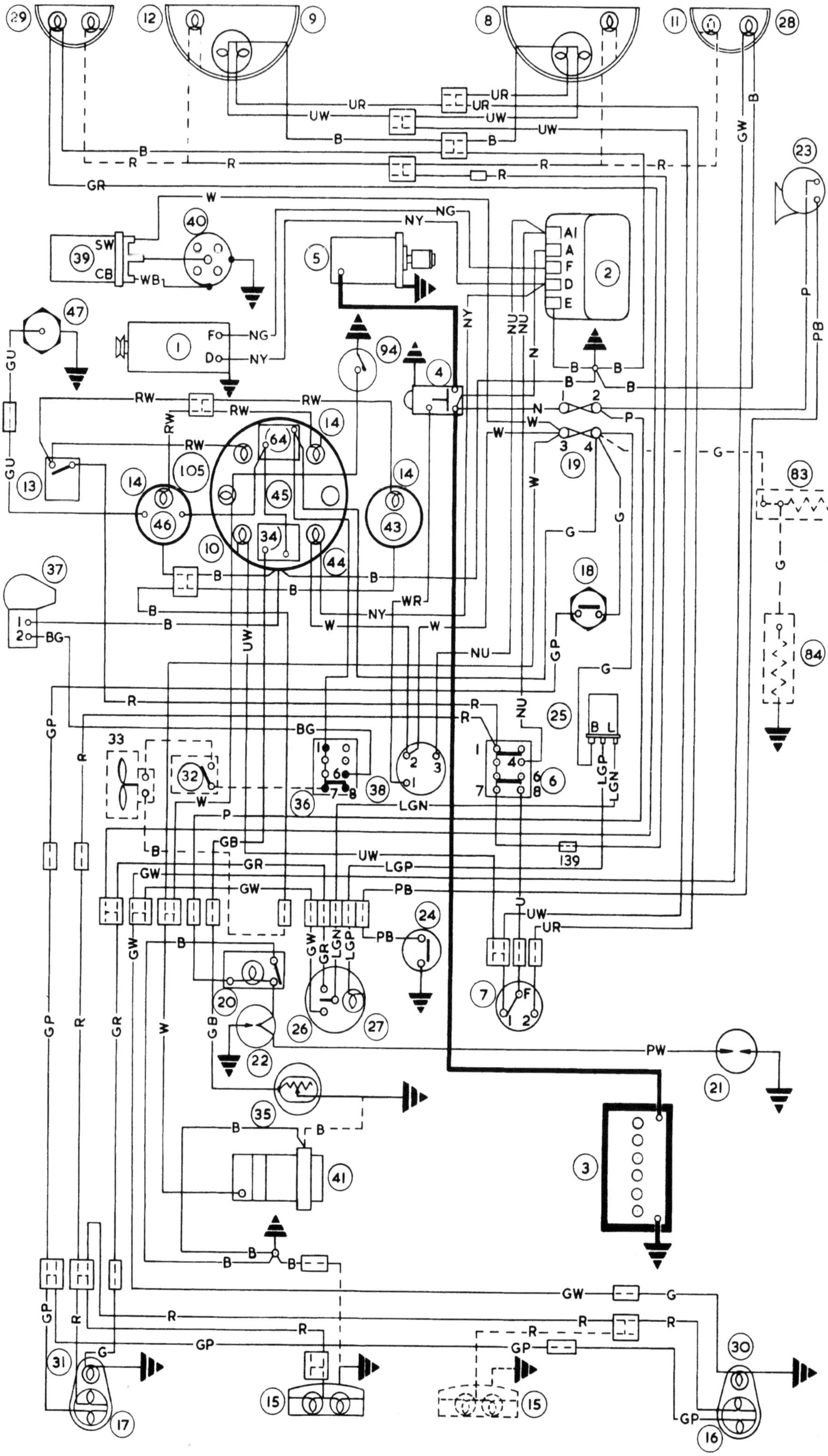

Figure 8.1(a) Conventional car wiring diagram

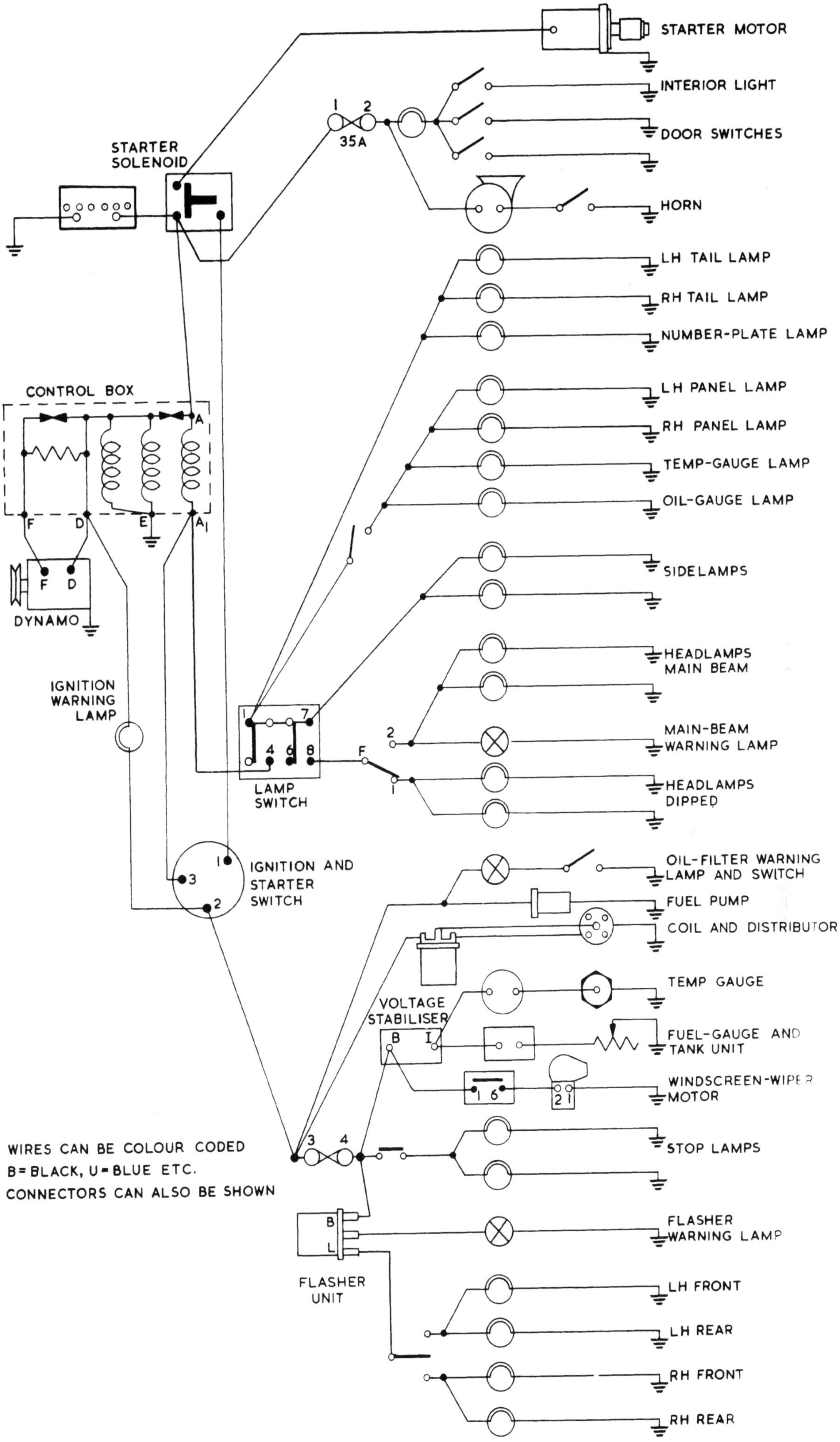

Figure 8.1(b) Functional car wiring diagram

(Reproduced by permission of Her Majesty's Stationery Office)

It was apparent from the start that one of the major factors influencing a man's performance is the set of aids with which he is supplied, the main aid being, in the case of radar equipment, circuit diagrams. It was also apparent that existing diagrams did not provide information in a form which was easily used. All the diagrams for the two equipments were therefore re-drawn in a form which was suitable both as an aid to fault diagnosis, and as a basis for training. It was this demonstration of the significant changes in diagrams which could be brought about by the application of simple functional rules, that gave rise to a major effort to provide diagrams which are considerably more useful for the maintenance task than those which were used in the past.

DESCRIPTION OF A FUNCTIONAL DIAGRAM

The accepted definition of the term functional diagram, derived from the work outlined on page 4/ is as follows:

'A functional diagram illustrates, in a simple form, one or more functions of a system (or part of a system). It emphasizes the flow paths of the signal (or comparable function) irrespective of spatial relationships. The term "functional" may be applied to any type of diagram drawn to satisfy this definition; thus: functional circuit diagram; functional wiring diagram; functional systems diagram.'

This definition omits one major characteristic of a functional diagram which is covered, in the standard of which the above definition will be part, by a general statement concerning the preparation of electronic diagrams. This is that the main sequence of cause-to-effect is from left to right.

As an example consider the car wiring diagrams of Figure 8.1. Figure 8.1(a) illustrates a typical car wiring diagram of the type found in almost every maintenance manual. The disadvantages of this diagram are that there is no systematic layout of symbols and that there are excessive cross-overs and parallel cable runs. Some attempt has been made to group the components according to their location in the car, but this is not consistent. Figure 8.1(b) contains the same information but is in functional form. The two main inputs are on the left and the outputs are listed on the right. The layout of the diagram has been arranged to simplify the illustration of wiring and emphasize the relationships between the components. In doing this the interpretation of a fault symptom is made much easier, as is a student's task of understanding the functions of the circuit.

Since Figures 8.1(b) and 8.2(b) illustrate primarily a change in diagram structure, the same symbols have been used in them as were used in Figures 8.1(a) and 8.2(a), and these diagrams are not, therefore, good examples of the use of British Standard symbols.

One point which is not illustrated very clearly in Figure 8.1 is that, in simplifying functional connections, unit boundaries can be drawn in any shape and in any position on the diagram; in other words the functional layout is in no way restricted by physical layout or boundaries. This point is illustrated in Figure 5.2 which shows the MOUNTING and I.F. AMPLIFIER boundaries drawn in an irregular shape to avoid distorting the flow paths. The unit boundaries are also drawn in several separate parts of the diagram in order to prevent long lines which represent unit boundaries from distorting the diagram. As an example, the MOUNTING is separately illustrated in squares A1, J3 and H4.

RULES FOR DRAWING FUNCTIONAL DIAGRAMS

The definition of the term **functional diagram** only provides a basis for drawing functional diagrams and it is possible to expand the definition into a set of rules which more fully set out the characteristics of such diagrams.

The rules for producing functional diagrams are:

a. The main signal flow must be from left to right.

b. Signal flow must be emphasized. This can be done by showing flow paths in a straight line, by avoiding crossover of flow paths, by avoiding closely spaced parallel paths and by using a thick line for the main flow path. Sloping lines may also be used for emphasis but should be used sparingly.

c. Arrows may be used to indicate direction of flow but must be used sparingly, and preferably only to resolve ambiguity.

d. Main inputs must be shown on the left of the diagram and main outputs on the right. (This is a direct result of applying rule (a)). Subsidiary inputs and outputs must be in clear spaces. The source of all inputs and the destination of all outputs must be shown.

e. Unit boundary lines may be any shape which does not distort the flow path. Units may be shown in more than one position on the diagram.

Additionally all diagrams must be drawn according to the following rules:

a. All information required for the task for which the diagram is provided must be shown on, or supplied with, the diagram. For fault location all plugs, sockets, controls, test points and terminals must be shown and referenced; test points at inputs and outputs are of particular importance. Normal state measurements for all test points must be available.

b. Symbols must be in accordance with the appropriate current British Standard. The relevant standards are BS 3939:1966 (Graphical Symbols for Electrical Power, Telecommunications and Electronics Diagrams), BS 108:1951 (Graphical Symbols for General Electrical Purposes), and BS 530:1948 (Graphical Symbols for Telecommunications) with Supplements Nos. 1-7. BS 3939 is an amalgamation and revision of the other two standaı ds, and where a symbol is specified in BS 3939 it supersedes its equivalent in the other standards.Figure 5.2 is an example of a diagram using the correct standard symbols. It is fully described in Chapter 5.

In functional diagrams some old drawing conventions must be abandoned. One of these is that all lines must be vertical or horizontal. This is convenient for the draughtsman, but not satisfactory for the user; sloping, and even curved, lines should be used if they are felt to be needed and add to the clarity of the diagram.

The drawing of diagrams which are closely related to the task for which they are required has the built-in advantage that it ensures that the required information is available. In the past it has often been the case that a diagram which is apparently full of information does not contain relevant information. On many occasions when diagrams have been converted to functional form it has been found necessary to refer to the actual equipment to obtain data which are not included in the old diagrams.

For large equipments more than one diagram is usually required to illustrate the complete structure. Mention must therefore be made of the need to prepare diagrams in such a form that it is easy to move from one diagram to another. Two points are important here. First, diagrams must be related to stages of the task, so that at one stage, say isolating a faulty equipment in a system, one diagram is used, and at another stage, say isolating a faulty unit in an equipment, a different diagram is used. Secondly, the transfer between diagrams must be easily made. This is the reason for the rule concerning the specification of signals at inputs to, and outputs from, the diagram. The user must be able to make a check at a test point shown near the input or output of a diagram and, depending on the result, be directed easily to the next diagram which is required. The overlap should be sufficient to confirm that the correct change has been made, but not so much that it confuses the user. The basic rule of left-to-right flow should ensure that the position of the appropriate flow path is located quickly.

EXAMPLES OF FUNCTIONAL DIAGRAMS

We will now see how the rules can be applied to produce functional diagrams, and we will use Figure 5.2 as an illustration. The main signal flow from aerial to headphones is left-to-right. It is emphasized by drawing in a horizontal line, and by using a thicker line than is used elsewhere. The meter in the control unit is also treated as part of the main path so that the main path diverges at the second i.f. amplifier. At no point does the main path cross another path and there are no confusing parallel paths adjacent to it. Subsidiary paths are also drawn using horizontal lines wherever possible with some sloping lines for emphasis. Arrows indicate the direction of flow where there may be some doubt, e.g. in squares 4A and 4B. The fact that unit boundary lines are not restricted to regular shapes has already been mentioned, so has the fact that the diagram uses current British Standard symbols. Plug and socket references, which are required to locate a faulty circuit, are shown on the diagram. Other information required for maintenance, such as test point voltages and waveforms, would be provided in the equipment manual, of which Figure 5.2 would be an extract.

Figure 8.2 shows how a circuit diagram can be drawn in functional form. Figure 8.2(a) shows a circuit diagram as it appears in an equipment manual. Although it is accurate and neat it does not present information in the best manner. For instance, it is not possible to decide quickly the function of the circuit, nor is it possible to see which circuit structures are involved. Figure 8.2(b) shows the same circuit drawn in functional form. Knowing that inputs are on the left and outputs on the right we can easily see that the function of the circuit is to derive one output from three inputs, i.e. it is a convergent circuit. The signal flow is strictly left-to-right, and it is drawn as simply as possible, avoiding crossover paths.

In circuit diagrams, a rule which further improves clarity is that components associated with a particular function must be grouped together. Thus, in Figure 8.2, although there are few components, those associated with each valve are grouped around that valve.

The car wiring diagram of Figure 8.1 is a good example of the great improvement which can be gained by placing inputs and outputs in specific positions, and as far as possible avoiding crossovers and parallel cable runs.

It is important to note that, in functional diagrams, the circuit structures discussed in Chapter 6 are clearly revealed. For instance, Figure 8.1(b) shows a predominantly divergent structure, and in Figure 5.2 we can easily see the divergent outputs (D3), and the convergent points at the mixer (B3)

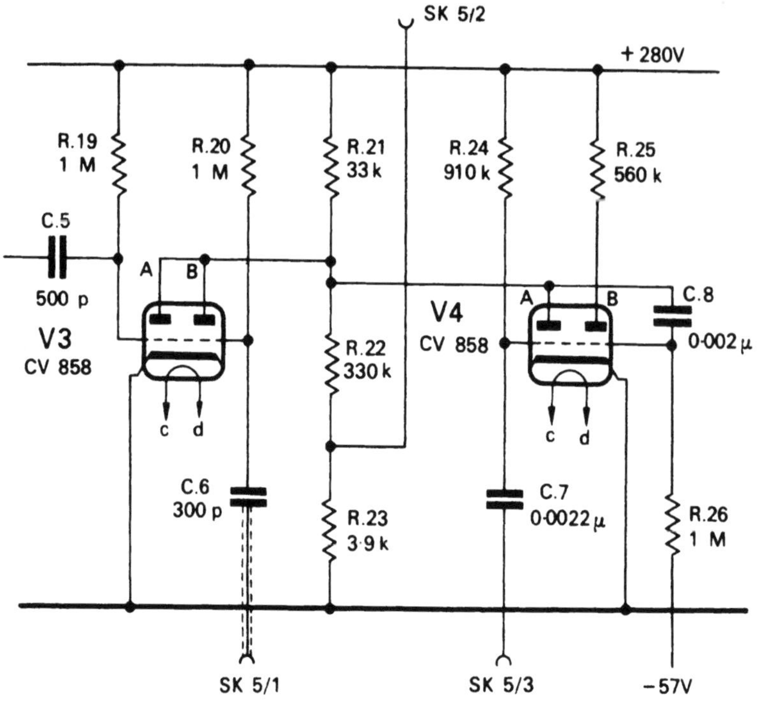

Figure 8.2(a) A coincidence circuit

(Reproduced by permission of Her Majesty's Stationery Office)

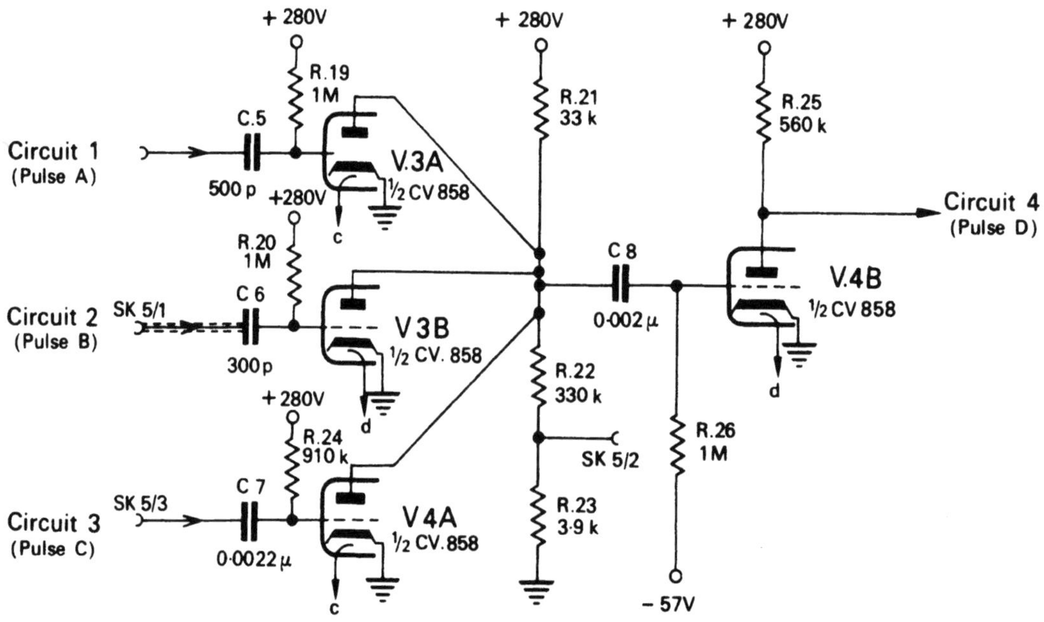

Figure 8.2(b) A coincidence circuit

(Reproduced by permission of Her Majesty's Stationery Office) and the first i.f. amplifier (C3). The fact that these structures can be readily identified results in the diagrams being good aids to fault location.

We have seen how several types of diagram can be drawn in functional form to improve their usefulness as an aid to fault location. Providing a device has some structure, and there is a flow of information, then it can be illustrated in functional form. Clearly we could not draw a functional diagram to illustrate a bag of marbles! In addition to electrical and electronic circuits, functional diagrams have been drawn for mechanical systems, fuel flow, air flow and water cooling systems, to quote a few examples.

USE OF FUNCTIONAL DIAGRAMS

Functional diagrams have been discussed in terms of their use as an aid to fault location. This was the reason for their development, and it is their main use. Naturally, they are suitable as an aid to any task which requires information to be illustrated in the same form as it is for fault location. Teaching the function of an equipment is an example.

By simplifying the structure of an equipment, functional diagrams reveal the types of circuit discussed in Chapter 6. Whereas, in an old type of diagram, it might be difficult to determine the inputs to a point of convergence, for example, or the components included in a feedback loop, with functional diagrams these are clearly illustrated in a recognizable form.

As an example of the use of a functional diagram for fault diagnosis consider again Figure 8.1. Suppose the engine starts but the horn does not work. What should be the next step towards locating the fault? A look at the functional diagram, Figure 8.1(b), tells us that the horn circuit includes divergent points at terminal 2 of the fuse, and at one of the starter solenoid contacts. Using the rule for divergent circuits we can deduce that the fault must lie in the circuit between the divergent point on the starter solenoid and the horn output.Since this area contains another divergent point at terminal 2 of the fuse we can again apply the rule and check one of the other outputs. The interior light is convenient: let us suppose that it does not work. The fault must then lie in the area common to the faulty outputs, which is between the starter solenoid terminal and terminal 2 of the fuse. In this small area the fuse would probably be checked first on the basis of relative reliabilities of the fuse. and the wiring.

It is worth considering the steps necessary to make similar deductions using Figure 8.1(a) instead of Figure 8.1(b). In the first place the horn would have to be located on the diagram. Its source of power would then have to be traced and a further check decided upon. None of these steps are easy, and a considerable amount of time would be necessary to carry them out. The technician using the diagram would probably be forced to adopt an inefficient method to locate the fault. It is certainly not clear from Figure 8.1(a) that a horn fault can quickly be localized by noting that the engine starts and the interior light does not function correctly.

The reader can assume numerous other combinations of faulty outputs, and make similar deductions to locate the fault. Since there are so many outputs the example gives good practice in the application of the rule for divergent circuits.

As another example consider Figure 5.2. Suppose that the meter output (J5) is correct, and that the earphone output is correct on REC but incorrect on ADF. Using the rule for divergence the fault must lie between the point at which the meter and earphone paths diverge (D3), and the earphone output. Using the rule for switching circuits the fault must also lie in the circuit which is only used when switched to ADF. This area comprises the moving contacts of the switch (G3) and the small circuit between the 15KΩ resistor (F3) and the ADF contact of the switch. The 15KΩ resistor itself is not suspect because it is part of the path to earth from SKX9, which is operative in the REC position. A good diagram has thus enabled us to reduce the possible fault source to a very small area using only a broad symptom pattern. The task would have presented much more difficulty had the circuit been illustrated in some other form, such as showing each unit on a separate page, which is often encountered.

CONCLUSION

Functional diagrams are written aids to maintenance which emphasize the structure of an equipment. They are suitable for all tasks which require knowledge of functional structure, the main one being fault location. Functional diagrams do not eliminate the need for diagrams which are intended for tasks such as manufacture or wiring.

Functional diagrams in the form described here are new and the technician will inevitably meet many older diagrams which are not drawn in functional form. It is important, however, that he should know what constitutes a good diagram, and that he should be able to visualize an equipment in functional form even if its diagram is not drawn in that form.

A logical extension of the functional principle is that tasks could be made much simpler if the construction of equipment were such that the mechanical layout corresponded to the functional layout; in other words, the diagrams as described in this chapter would also serve as layout diagrams. Unfortunately this is a concept which, apart from very simple circuits, is unlikely to materialize. The functional diagram will, in general, bear little resemblance to the mechanical structure of the equipment that it illustrates.

SUMMARY

Functional diagrams illustrate in a simple form the function of a system; they emphasize flow paths of the signal irrespective of spatial relationships.

The main rules for drawing functional diagrams are:

1. The main signal flow must be from left to right.

2. Signal flow must be emphasized e.g. by arrows, or thick lines.

3. Main inputs must be shown on the left and main outputs on the right; subsidiary inputs and outputs must connect in clear spaces.

4. Unit boundary lines may be drawn in any shape, and in any position, that does not distort the flow path.

5. Information concerning test points, plugs, sockets and terminals which are required for maintenance must be shown on the diagram.

6. Symbols must conform to current British Standards.

The use of functional diagrams makes maintenance, and in particular, fault diagnosis, much simpler.

CHAPTER 9 FAULT-FINDING GUIDES

INTRODUCTION

A fault-finding guide is a maintenance aid, which, as the name implies, provides guidance for detection and location of faults, in the non-routine maintenance task. Of course, the nature of the guide will depend on the method of location which is most appropriate to the equipment concerned. This may be any one, or a combination of, the methods described in Chapter 5, but, since the commonest methods are systematic ones based on either functional structure or reliability data, we will concentrate on these. In particular, guides which are based on the half-split method will be explained.

Because most guides cover both detection and location of faults, we can split them into two parts which we will call the **functional check**, which covers fault detection, and **fault location notes**, covering the location aspect.

This chapter describes guides for functional checks, and also the main types of fault-location notes, based on the half-split and the reliability data methods of fault location. These are often presented as tree charts and symptom-cause tables respectively.

FUNCTIONAL CHECKS

When he is faced with an equipment which has been reported faulty and in which he has to locate a fault, the technician must take a series of steps in order to provide information upon which to base his initial deductions. In simple cases this would be merely operating the equipment, for example switching on and turning up a gain control; but for more complex equipment it could consist of a large number of actions. In addition to operating the equipment the technician will want to collect important fault symptoms, and items must be incorporated in the check for this purpose.

A functional check is a series of actions and checks which are to be carried out with the intention of providing fault information. Such a check will not be as detailed as the routine maintenance schedules which were discussed earlier in this book, but will be more detailed than the operating procedure for the equipment. How exactly a particular functional check falls between these two extremes will depend on the balance between the information which is obtained from checks, and the time taken to complete them. One thing, however, must be certain: if a functional check is completed and no apparent fault symptom appears then, as far as the operator is concerned, the equipment does not contain a fault.

The order in which items are listed in a functional check depends on the requirements for operating the equipment, and on the method of fault location to be used. For instance, if an equipment requires that heaters are switched on before an h.t. supply, then this must be the order which is used, irrespective of any other considerations. Any equipment with a safety interlock system will have such requirements. The method of fault location governs the order of checks when there are no overriding operating checks. If the half-split method is being used, then each check should be such that the faulty area is halved whatever the result of the check; if the reliability data method is used, then checks will successively eliminate the most unreliable units or components.

Table 9.1: Example of a Functional Check

ITEM	ACTION	EXPECTED RESULT	SEE NOTE
1	Switch on 115V 1,600 Hz bench supply.	A.C. meter shows 115V. Valves light up in R unit and radar head.	1
2	Turn AUTO/MAN switch to AUTO.	A.C. meter remains at 115V.	2
3	Observe oscilloscope S.	A spot with, or without, wings. (If no spot is visible attempt to obtain one by adjustment of AZ SHIFT, EL SHIFT, AMPLITUDE and BRIGHTNESS controls).	3
4	Adjust BRIGHTNESS control on control unit 116.	Brightness of spot varies.	4

ITEM	ACTION	EXPECTED RESULT	SEE NOTE
5	Adjust FOCUS control on control unit 116.	The focus of the spot varies and it is possible to obtain a clearly defined spot.	5
-			
-			
-			
-			
10	Ensure that the scanner is not obstructed.	-	-
11	Switch on the +28V supply.	-	-
12	Turn the scanner ON/OFF switch to ON.	The scanner rotates.	9
13	Observe oscilloscope S, adjusting BRIGHTNESS if necessary.	A raster of horizontal lines should be displayed.	10
14	Adjust AZ SHIFT and EL SHIFT controls.	It should be possible to obtain a raster just filling the rectangle marked on the c.r.t. face.	11
15	Observe oscilloscope R, adjusting BRIGHTNESS if necessary.	A raster of vertical lines should be displayed.	12
16	Adjust R AZ SHIFT and GAIN controls.	It should be possible to obtain a display with edges just visible through the holes provided in the c.r.t. case.	13

An example of a functional check for a radar equipment is shown in Table 9.1. It is written in four columns. The first gives the item number, the second shows the action to be taken, the third shows the expected result of the action under normal operating conditions, and the fourth column gives direction to notes on fault location, i.e. what to do if the expected result is not obtained. Two extracts from the check are shown; in the top extract (items 1 to 5 inclusive) the order is governed by operating procedure. The equipment has to be switched on, and the cathode ray tube brightness control turned up before other symptoms can be detected; the bottom extract (items 10 to 16 inclusive) is governed by the rule for divergent circuits. The timebase waveform for both R and S displays originates from the same divergent circuit, and the outputs are checked in turn according to the rule for divergence.

In most cases it is essential to carry out the items of a functional check in the order specified, because only then will the fault-finding notes be accurate. Sometimes, however, provision is made for entry at particular points in the guide, depending on what information is already available about the equipment e.g. from an operator's report. In these cases items related to switching on will need to be completed before any location items.

FAULT LOCATION NOTES

If the functional check is completed with no incorrect results then the equipment is functioning correctly. If any item gives an incorrect result, then a fault has been detected and a fault location procedure must be followed. As already mentioned the actual procedure will depend on the method of fault location to be adopted. We will discuss notes which are based on the half-split method, and notes based on the reliability data method.

NOTES BASED ON THE HALF-SPLIT METHOD

Since the half-split is a sequential method, notes relating to it will show a sequence of checks and possible results. Each check will have at least two possible results – correct or incorrect – so that the sequence can be illustrated by a tree, with its top representing the initial symptoms and its roots all the possible faulty units or components. A simple example of this form of notes commonly known as a **tree chart** is shown in Figure 9.1. The original symptom may, of course, appear as the result of one item in the functional check, or it could be a combination of incorrect and correct results which are obtained by completing the functional check before starting fault location. The latter is more common, and generally indicates a sounder start to the procedure.

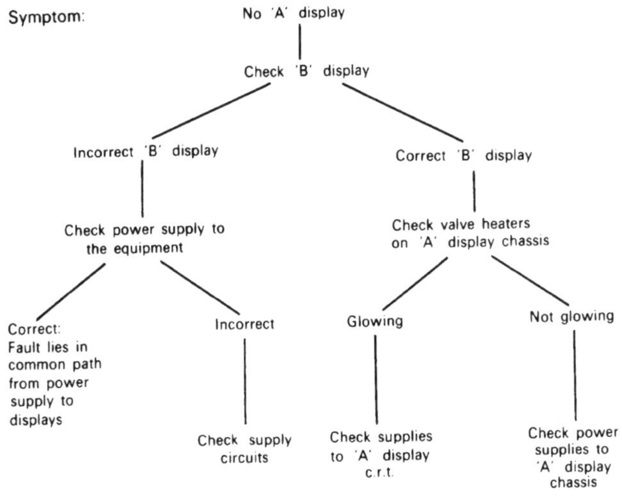

Figure 9.1 Example of a tree chart

Each step must be based on the diagnostic rules, but in practice account is taken of the ease with which a check can be made as well as of the theoretically best place to make a check. Figure 9.2 is an example of fault-finding notes based on the half-split method. It refers to the practical series circuit of Figure 9.3, in which one of the units contains a fault. All checks divide the possible sources of the fault in half or as nearly as possible in half, so that if the notes are followed correctly, a fault anywhere in the circuit can be located using an efficient series of checks.

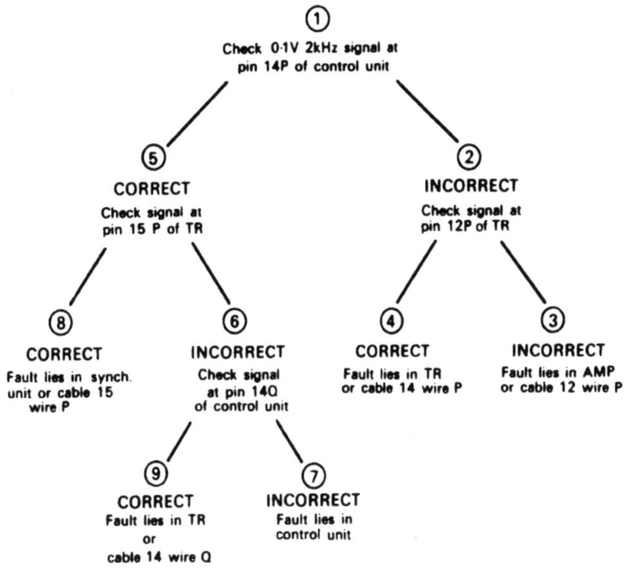

Figure 9.2 Fault-finding notes based on the half-split method for use with Figure 9.3

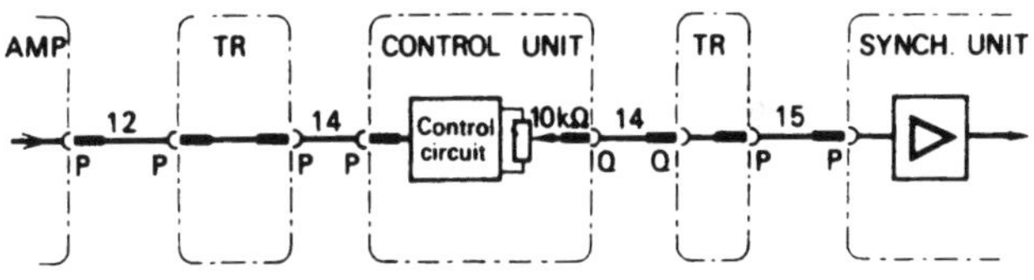

Figure 9.3 A practical series circuit

Naturally, if conditions occur in which the rules do not apply, and the notes have been written only for rigid application of the rules, there will be difficulty in using them. This could occur for the difficult types of fault discussed in section 6.12. It can be overcome either by providing the technician with functional diagrams and training him to use them for locating difficult faults or by arranging that assistance can be given to the maintenance staff when such faults arise.

Criticism of the tree chart is often made on the grounds that, to make it sufficiently comprehensive, an impossibly broad base is required containing a very large number of roots. This criticism is fully justified if the chart is taken as far as component level for all symptoms, because the base of the tree will consist at least of a list of all the components in the equipment. However, this difficulty arises with all forms of fault location guide. It is overcome if we bear in mind that the guide is acting only as an aid, and that an aid is only provided where it is necessary for efficient work. Thus a guide should aid the technician with the most difficult part of the task, but leave him to take over at a point where training and experience have provided the necessary skill.

The guide of Figure 9.2 takes the user to a stage where there are no more possible faulty areas than one unit and a cable. It could have gone further to isolate the fault conclusively to a unit, or it could be argued that if the user has been taught the half-split rule he should not require the guide at all. It follows that guides will be of most use to technicians with the least training and least experience. Although this is perfectly true, it should not deter even the most experienced technician from using them. In the long run, following the specified fault location procedure will pay dividends.

The roots of a tree chart based on the half-split method will thus occur where the structure of the circuit is such that the technician has been taught how to deal with it, e.g. a linear or divergent structure.

Good fault finding guides are characterized by the fact that the questions posed at each stage have precise answers. The question: 'What is the display state?' permits a great number of answers, but, 'Does the display show a 50-Hz sine wave of 10V amplitude?' can be answered specifically by 'YES' or 'NO'. All check sequences can be split down to lead to clear 'YES' or 'NO' answers.

OTHER METHODS OF PRESENTATION

The layout of a tree chart as illustrated in Figures 9.1 and 9.2 is not the only way of presenting the information, and it is not necessarily the best way. In this Chapter we will only discuss written methods of presentation, but it is possible to write the guide so that it can easily be used in some other form of presentation. With this in mind the tree chart could be constructed so that **all** questions require only 'YES' or 'NO' answers, so that all 'NO' answers are illustrated in one particular position and all 'YES' answers in a different position. Using Figure 9.2 as an example, the checks and answers could be re-positioned as shown on the chart of Figure 9.4(a) in which the numbered boxes correspond to the numbered checks on Figure 9.2. All the 'NO' answers are shown on the right of a question, and all the 'YES' answers below a question. A further simplification is to separate the text from the diagram, using the box numbers for cross reference. Thus in Figure 9.4(b) a correct answer to check 1 leads to check 5, and an incorrect answer to check 1 leads to check 2; this relationship is illustrated as boxes 1, 2 and 5 in Figure 9.4(a).

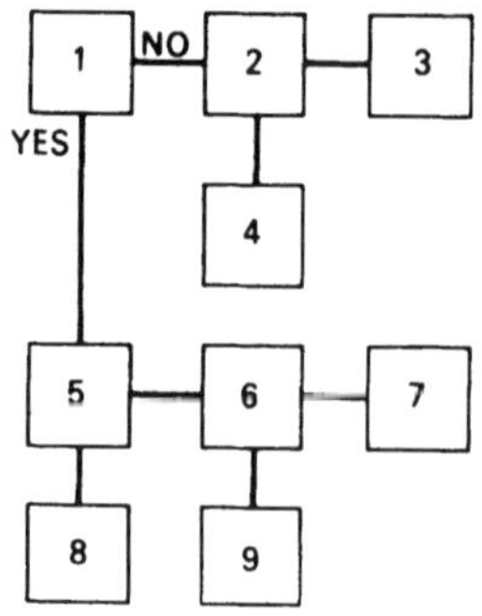

(a)

Check number	Action	Result	
		Correct	Incorrect
1	Check 0·1V 2kHz signal at pin 14P of control unit	5	2
2	Check signal at pin 12 P of TR.	4	3
3	Fault lies in AMP or cable 12 wire P		
4	Fault lies in TR or cable 14 wire P		
5	Check signal at pin 15 P of TR.	8	6
6	Check signal at pin 14Q of control unit	9	7
7	Fault lies in control unit		
8	Fault lies in synch. unit or cable 15 wire P		
9	Fault lies in TR or cable 14 wire Q		

(b)

Figure 9.4 Alternative representations of Figure 9.2

The disadvantage of adopting the presentation described is that it is not easy to relate a particular check to all other checks, or quickly to see which areas of the equipment have been eliminated by a particular check. The advantages are that the material is in a form which is suitable for automatic testing, that there is no restriction on the amount of detail which can be given in a check, and that both diagram and text are easily presented. In the conventional tree chart a lengthy check would cause considerable distortion of the structure.

A variation of the basic form of presentation is sometimes met when the checks require a long explanation, or when a lot of test data is provided. This consists of writing each check on one page of the guide, and referring to page numbers in the EXPECTED RESULT column. Thus, in following the guide, the user would be moving from one page to another, but not necessarily in the order in which they are printed.

The advantages of fault location notes, based on the half-split method, as an aid to maintenance are:

a. They are based on an efficient method of fault location.

b. They can be written to cover a maximum number of fault symptoms and causes.

c. Since they are based on functional structure, they can be prepared before an equipment comes into operation. This is not possible for guides requiring accurate reliability data.

d. Like all aids they affect the training requirement. Since the work is made easier by specifying all the decisions to be made in fault location, the amount of training required is reduced.

Some additional advantages accrue from the fact that the fault location task is clearly defined. These are:

a. It is easy to compare tasks, and it is thus easy to predict training requirements. This is because training will only be required where knowledge or skill which is additional to that already available becomes necessary.

b. It is easy to measure and compare the performances of men performing the task, and thus provide useful information for management.

c. It is easy to predict the time necessary for repair and thus to perform availability calculations. (See Section 2.9).

Although few fault location notes are at present issued in the formats shown in Figure 9.4, it is likely that they will become necessary because of the difficulty of providing the information in simple tree chart form.

NOTES BASED ON THE RELIABILITY DATA METHOD

A popular and easily prepared form of aid is a tabulation of possible symptoms and their most likely cause or causes. This is not strictly the systematic reliability method as described in Chapter 5, but is a combination of that method and the non-sequential method of moving directly from an observed symptom to a possible cause; hence the common name of **symptom-cause table** for the aid.

Symptom	Unit	Remedy
1. Raster too small, No Rx signals	Radar head	Replace R.160
2. No A.F.C. sweep No Rx signals	"	Replace R.105
3. Raster remains on Tx	"	Replace R.102
4. Tx only when switch held on Tx	"	Replace R.101
5. No bright-up	W.F.G. 76	Check C609, V605A, V507 Replace unserviceable component
6. Bright-up but no wings	"	Check C810, V805. Replace unserviceable component
7. No warning note	"	Replace V604
8. Distorted wings display	"	Replace V807

Figure 9.5 Example of a symptom — cause tabulation for a radar receiver
(Reproduced by permission of Her Majesty's Stationery Office)

Symptom	Possible cause
No picture, no sound	Setting of channel selector switch; fine tuning;SKI; PLI; VI; V2
Sound correct, no picture	Ion trap magnet; V3; V5; V7; V10; CD1
Pale picture	Fine tuning; contrast and brilliance settings; CD1; L24
Flat highlights	Setting of vision noise limiter plug; V7B; V10
Poor definition	Fine tuning; V7; V10; R51; C65; C67 setting; L24
Excessive width	L27A open circuit; L28 short circuit

Figure 9.6 Example of a symptom — cause tabulation for a television receiver

Examples of symptom-cause tabulations are shown in Figures 9.5 and 9.6. Figure 9.5 refers to a radar receiver and Figure 9.6 to a television receiver. All the user has to do is to match the symptoms obtained with those in the notes and check the indicated components.

The advantages of this type of guide are that it is very easy to use and, when successful, provides a quick result. If only one cause is given for a particular symptom, as is the case with the first symptom in Figure 9.5, the process of systematic elimination is not required; that particular part of the table uses a non-sequential method of location. If, on the other hand, a number of possible causes are given, as for symptom 5 of Figure 9.5, these have to be systematically eliminated. This is usually done on the basis of a compromise between the ease of doing the check and the component reliabilities. For symptom 5 V605 and V507 would, therefore, be checked, probably by replacement, before C609, since on the average valves are less reliable than capacitors and are usually more easily exchanged.

There are two main disadvantages of the symptom-cause tabulation. First, it relies on accurate information being available concerning the reliability of components; this is because, to be effective, the probability that a particular fault will produce the given symptom must be high. Chapters 2 and 3 discussed reliability in detail, and the difficulty of forecasting component reliabilities was stressed. Even if all component reliabilities can be accurately forecast, it is unlikely that a design would be such as to include a particular component or components whose reliability is very much worse than others. Forecasting reliabilities is therefore unlikely to provide a very accurate symptom-cause tabulation. A much better way of obtaining component reliabilities for an equipment is from records of fault occurrences in the equipment. In this way an accurate symptom-cause chart could be written, but it would not be available for use when most required, that is in the early stages of an equipment's life; nor would it remain accurate for very long because one would expect modifications to be made to an equipment to eliminate major sources of unreliability.

The second disadvantage is that the causes listed for each symptom can only be a few of all the possible causes, so that the guide is bound to fail on some occasions. When it does fail the user is left with no aid at all to help in his task, and he must then produce his own method and checks to locate the fault.

Unlike the tree chart, the symptom-cause tabulation does not provide the sort of information that can be easily used in automatic fault location devices. This is mainly because the tabulation has no rigid and recognizable logical structure.

THE USE OF FAULT-FINDING GUIDES

Fault finding guides have been described as written aids to the maintenance task. The technician will, no doubt, in the course of his work meet various methods of presentation, both written and automatic. The next Chapter describes under the heading 'Automatic Test Equipment', how guides can be presented automatically.

Whatever the content of a guide may be, and however well it may be presented, it is only an effective aid if it is used properly. Technicians should therefore be urged to use guides where they are available, and to follow closely the instructions in them. Human error in the use of the guide is unavoidable so it is worth repeating a series of checks before a guide is declared to be inapplicable for a particular task.

It is also worth carrying out double checks where they are possible and easy. When using notes based on the half-split method, for instance, the route to a fault can also be followed on the functional diagram for the equipment to provide a rough check that a reasonable series of steps is being taken.

SUMMARY

Fault-finding guides provide guidance for the detection and location of faults; the most common type of guide is written, although aural, visual or automatic guides are possible.

The detection of faults is carried out by means of a functional check, which is a series of actions and checks which are intended to provide fault information. A written guide covering a functional check is shown in Table 9.1.

The location of faults is mainly carried out by the half-split or reliability data method described in Chapter 5, and written fault location notes can be used as guides to help technicians in fault tracing.

Fault-location notes based on the half-split method are usually shown as tree charts, of which an example is shown in Figure 9.2. Tree charts are efficient, have wide coverage of faults and can be prepared before an equipment is brought into operation.

Fault-location notes based on the reliability-data method are usually shown as symptom-cause tables, an example of which is in Figure 9.5. They are easy to use, but not always successful.

CHAPTER 10 TEST EQUIPMENT

INTRODUCTION

Test equipment is the means by which the performance and other characteristics of an equipment are checked for the purposes of routine or non-routine maintenance. It enables the technician to collect necessary data about an equipment such as voltages, waveforms and circuit characteristics, in order to compare them with expected figures.

In discussing test equipment it is convenient to refer to three categories; general, special and automatic test equipment. General test equipment is that which is used in a wide variety of tests and equipments; special test equipment is used in only one test and on one equipment, or in a very limited number of tests; automatic test equipment has provision for part of the testing process to be carried out without the need for an operator.

There are many books dealing with the laboratory use of test equipments, which give details of their circuitry and how to use them. It is not intended to duplicate this detailed information in this chapter, but only to discuss test equipment in general terms. Our object in this chapter is to clarify the role of different types of test equipment in the maintenance system.

THE USE OF TEST EQUIPMENT

Here we will mention only a few points which are of particular importance and often ignored by the technician. First, let us consider the question of tolerances. There are three aspects: the amount of variation which is permissible in a particular quantity; the effect of connecting an instrument to measure the quantity; and the accuracy of the instrument used. The problem is simplified when a checking instruction specifies not only where the check should be made, but also how it should be made and the permissible tolerance. For instance, an instruction which says 'Measure the voltage at pin 4 of V806 using a model AB multimeter; the reading should be 8V ± 1V', takes into account the effect of measurement and the permissible tolerances. It says, in fact, that if this reading is obtained on the meter the voltage at pin 4 is correct. It does not necessarily say that the voltage at pin 4 is that which is read on the meter; meters seldom display the actual quantity applied to their terminals.

Difficulty arises when a check is not specified so precisely, or is not specified at all. It is then that the technician must bear in mind the three aspects mentioned. He must make some realistic assessment of the permitted variation in the quantity being measured, he must ensure that connection of an instrument has only a small effect on the quantity being measured and he must also make some assessment of the accuracy of the instrument. It is therefore usually recommended that some form of cross-check on a reading is made, that meter readings are kept as near full scale deflection as possible, that no attempt is made to read a scale more precisely than to the smallest printed scale division, and that instruments are regularly checked against a standard. A point to note is that meter accuracy is often specified as a percentage of the full-scale deflection and that this remains fixed over the whole meter scale. Thus an accuracy specified as ± 2% of full-scale gives an accuracy of only ± 20% for a reading which is one-tenth full scale. This is the reason for advice to keep the meter reading near full scale deflection.

Digital instruments are becoming more and more popular. They have the advantage that accuracy is not reduced by interpolation and that there is no need to align a meter needle with a printed scale.

The necessity for taking care of instruments and taking safety precautions in their use cannot be over-stressed. Although modern instruments are generally more robust than older types, mainly due to their use of semiconductor devices, they are also required to measure a much wider range of quantities; measurements of frequencies of hundreds of megahertz, currents down to fractions of a nanoampere and voltages down to fractions of a microvolt are commonplace. Special care is therefore always necessary in handling and use of such measuring instruments. An instrument's earthing, fusing and mains tapping arrangements should be noted before use; precautions against overload, safety precautions, correct ventilation and mechanical stability are important in use; and protective measures are often necessary when it is out of use, such as not leaving an oscilloscope beam focussed on one spot and shorting the input terminals of sensitive meters.

The very nature of testing requires that access be made to the working parts of an equipment, and thus the risk of danger is increased. In dealing with any live equipment it is sound practice to keep one hand in the pocket, to stand on a good insulating surface and to move wires with the back of the fingers to avoid gripping the wire involuntarily if a shock is felt. Handling live equipment should be avoided if possible, covers should not be removed unless it is absolutely necessary and safety switches should not be by-passed. A point which is often forgotten is that when leads are connected from the equipment to an instrument, the instrument terminals are live and at the same voltage as the equipment test point.

GENERAL TEST EQUIPMENT

General test equipment has facilities for a wide range of measurements on a wide range of equipments. Clearly, the more versatile is a piece of equipment, the more likely it is to be in constant use, and it becomes an economical investment. The user may have to buy only one such instrument where otherwise it would be necessary to buy several. There are also advantages in the use of general test equipment from the training point of view; a person who has been trained to use a universal piece of test equipment need not be re-trained when he moves to another job, providing the same, or a similar equipment is available.

The main types of general test equipment, which can be found in almost all laboratories, are: multimeter, electronic voltmeter, oscilloscope, d.c. power supply, a.f. signal generator, r.f. signal generator and transistor or valve tester. The first three – multimeter, electronic voltmeter and oscilloscope – are used for signal detection, i.e. to measure the value of the signal at a particular point in the equipment. The power supply and signal generators are for signal injection, i.e. to energize an equipment and to provide the sort of stimulus it would receive in normal operation. Signal injection can be used in two different ways. It can merely provide a signal to an equipment at the normal input point to enable signals to be detected at appropriate test points, or it can provide an input to one of a number of points throughout the equipment, so that the normal equipment outputs only are used as the means of detection. In discussing methods of fault location earlier in this book, emphasis was placed on the detection of signals at appropriate test points, but it must be emphasized that all the methods could have been described in terms of signal injection at appropriate points, and the same results would have been obtained. The last item on the list, the transistor or valve tester, is a device which measures the state of a particular type of component so that the user can determine whether it is fit for use. The multimeter is, of course, able to measure the state of some of the other common components, or some of their characteristics, for example the resistance of an inductor.

Manufacturers of general test equipment emphasize its versatility and do their utmost to extend its range of use, sometimes by the provision of accessories. Although this is a fully justifiable aim, the user must remember that any one piece of equipment can only be in one place at a time. The advantages of providing a small number of very versatile instruments could, therefore, be partially offset by lack of availability of the instruments.

SPECIAL TEST EQUIPMENT

Special test equipment is required to inject, or detect, voltages or signals which the general equipment cannot cope with. The commonest requirement is for complex waveforms which cannot be provided by conventional signal generators and sometimes cannot be detected easily by the conventional oscilloscope.

The fact that the use of special test equipment is restricted means that it is essentially part of the main equipment. In many cases it is permanently built in to the main equipment. Its disadvantages are the converse of the advantages of general test equipment: it does not have wide applications and thus has limited use; and it requires to be learnt with the equipment because the technician is likely to require different knowledge and skills for each item of special test equipment. Usually every effort is made to eliminate the need for special test equipment or to extend its use to at least a narrow range of equipments.

The advantage of special test equipment, particularly if it is built in to the main equipment, is that it provides rapid and accurate test data with the minimum chance of errors being introduced by the user. Quite often a rotary switch will enable a dozen or so test points to be checked rapidly without the need for external connection or setting up. Of course, the technician still has to interpret the results, and he will probably have to continue a maintenance task by making additional checks to narrow down the area in which a fault lies.

AUTOMATIC TEST EQUIPMENT (A.T.E.)

With the ever-increasing complexity of modern electronic equipment, and the associated increase in maintenance requirements, it is not surprising that parts of the maintenance task have been automated; indeed for some aspects it has become essential to automate. The advantage of automation is mainly the saving of time. As an example, one aircraft test routine has been reduced from between 7 and 8 hours to only 23 minutes by the use of automatic testing. Other advantages are the improved precision and standardization of performance tests, and a simplification of the task which enables the work to be done by relatively unskilled operators. Although the actual operating task may be simplified, it must be remembered that someone has to prepare the test procedures whatever presentation device is used. A.T.E. can only carry out the instructions given to it by its designer.

The storage capacity of most automatic equipment is high, and improvements can be made in recording maintenance tasks so as to provide information on reliability, spares requirements and, if necessary, the performance of individuals doing the work.

One area in which automating routine maintenance has proved of great value is in-flight recording for aircraft. Small in-flight data recorders enable the performance of all parts of a complex equipment to be recorded during actual operation, a condition which is often extremely difficult to reproduce on the ground. Apart from the gains obtained in respect of maintenance, such data can be used for aircrew evaluation and training purposes. Because of the bulky equipment required, the data provided by recorders must be processed on the ground, and it is here that most difficulties arise, because a vast amount of data has to be analyzed to detect deterioration in performance, and to associate the deterioration with a particular unit or component. At present, the state of the art is such that although discrepancies or deterioration in performance are detected, subsequent action must still be decided by the user. A large number of civil and military aircraft have in-flight recorders installed in them, and facilities for processing the recorded data are increasing rapidly and becoming more versatile.

On the ground, the need for detailed recording of operational data is less because the maintenance technician is usually able to operate the equipment and reproduce fault symptoms; there is also an increasing tendency to reduce to a minimum the requirements for routine maintenance. Nevertheless, there are important areas in which the A.T.E. may be used on the ground equipment; one application is in checking the readiness state of aircraft and ground-based missiles. A.T.E. enables a series of checks to be carried out rapidly at optimum intervals to try to ensure that the equipment will perform its function when required. In addition the use of A.T.E. permits the technician to carry out the tests at a place remote from the equipment. Used in this manner A.T.E. reduces the number of systems which are required to ensure that a high proportion fulfill their function.

In discussing the use of A.T.E. for non-routine maintenance tasks we must consider the three phases of detection, location and rectification. In the hypothetical case of all three phases being automated the equipment will never go wrong and its reliability is effectively increased to an infinite m.t.b.f. This is clearly not possible, and it is found that A.T.E. is usually provided for the detection and part of the location task only, though in some cases temporary rectification is carried out. The human operator must then be included in the system, not only to operate the equipment, but also to complete the detection and carry out the rectification tasks. Since the range of A.T.E. is so wide, both in respect of the amount of the task it automates and of the manner in which it is done, we shall only describe some examples of the type of A.T.E. which is likely to come into greater use in the near future.

AUDIO EQUIPMENT

Devices which use magnetic tape as an information store, and earphones or loudspeaker as a means of presenting information, provide examples of relatively simple A.T.E. Instructions for carrying out checks, each devized so as to require only a 'YES' or 'NO' answer, are recorded serially on the magnetic tape. By operating a switch the user plays back over the earphones the instructions for one check; he carries it out and operates a switch to record the answer. The next check played back depends on the answer given; its location on the tape could be next in series on the same track, elsewhere on the same track, or on another track. Simple versions of such a device would provide only for the first and last of these. The simple device could give guidance for only a few steps in a fault location sequence, and it would be necessary for the information which is needed for long fault location sequences to be provided in some other form, such as a written tree chart. Part only of the fault location procedure could therefore be automated with a simple device.

More advanced versions of this type of device use one track for recording checking procedures and another (or others) for digitally coding each check. Depending on the result of a check, the device can be constructed so that the tape moves the required distance forward or backward to locate the next appropriate check. The number of checks in a series is therefore almost unlimited, and the device can be programmed to locate faults to the level of a replaceable part. In Chapter 9 we discussed a written aid to fault location which presents information in exactly the form which is required for an advanced audio device of this type.

The provision of an audio testing device makes the maintenance task easier by eliminating the need for the operator to decide what check to make next; he does, of course, still have to make checks and therefore requires the necessary skill for this aspect of the job. Audio devices also eliminate the sort of errors which stem from following a written test sequence, which, in most circumstances, would be the alternative means of presentation; they are more durable than written instructions, and in operation they leave both hands free for manual tasks.

A simple audio device of the type described has been in use for testing aircraft systems for a number of years, but the idea is by no means only applicable to aviation and there will undoubtedly be an increase in the use of both simple and advanced versions in the future.

TAPE-CONTROLLED TESTING

Another form of static or mobile A.T.E. which is in use for both civil and military aircraft automatically carries out pre-determined checks on a system, or sub-system, according to instructions provided on a tape input. The equipment contains high grade signal sources, the characteristics of which are controlled by relays set by the tape control. These signals are connected to the appropriate inputs by means of further tape-controlled relays. When the input has been set, the output from the required test point, converted into digital form, is compared with the normal state result and tolerance expected at that point. Results of all tests can be printed out, and a tolerance alarm system can be included to stop the printer and test procedure if an out-of-tolerance signal is detected.

At present this type of device is capable of carrying out single tests, by manual selection, and a number of test sequences which are determined by the available tapes. A variety of types of test, including measuring a.c. and d.c. voltages, frequency, phase angle and time intervals can be made, and stimuli in the form of a.c. and d.c. voltages, and steady state, ramp or step functions can be provided. In this type of A.T.E. there is no feedback of test results to the control system, so that the checks carried out are not dependent on the result of previous checks, and an efficient sequential method of fault location cannot be followed. These devices therefore automate the fault detection aspect of the task only and in doing so provide a collection of symptoms, in the form of correct and incorrect results, for the technician to use in fault location.

Like most A.T.E. the type of device described is made to be self-checking by means of built-in test facilities and a comprehensive self-checking procedure. One version automatically goes into a self-checking routine, to prove that it is the equipment and not the tester which has failed, whenever a fault is detected. Eighty percent of the equivalent manual testing time can be saved by using devices of the nature described.

A.T.E. FOR COMPUTERS

Computers lend themselves readily to automatic testing because of their flexibility of operation, and because input and output devices are provided as part of the equipment. A test sequence only requires that the appropriate instructions are fed in. With large computers, for which down-time is expensive, efficient maintenance is of prime importance.

One method of increasing the availability of digital computers, which cannot be used for other equipment, is the use of error-correcting codes. These can be designed to locate any number of incorrect digits in a word and correct them by changing them to the other state. The operation of the computer is thus uninterrupted. The simplest version of an error-locating code is the provision of two dimensional parity checks; for example, for data on tapes, additional digits are added to rows and columns to make them consistent in possessing an odd (or even) number of digits of a particular kind. An error exists where row and column inconsistencies coincide. Of course, when a fault is detected and corrected by means of a code, it is necessary to indicate the fact in some fashion so that the fault can be located at a more convenient time, say at the end of the current calculation. The provision of error-correcting codes adds to the hardware, and hence to the cost of a computer, but they can be justified for large computers where the interruption of a long process could be expensive.

Most computer fault diagnostic programmes contain provision for carrying out simple specified functions, recording the results, comparing these with the expected results and hence making deductions concerning the location of the fault. The amount of fault location which can be done automatically varies considerably. At one end of the scale, fault symptoms in a computer's central processor can be obtained by displaying all register contents and the state of other critical points in the logic. A word can be read in to the computer, and the processor stepped slowly to compare what actually happens with what should happen. This method calls for considerable care, and it still leaves the operator to make deductions concerning the source of the fault from the symptoms obtained. At the other end of the scale, the computer itself, or a satellite computer, could use the symptoms to isolate the fault to a single replaceable unit. Few programmes, however, are as specific as this.

In general computers require a number of fault diagnostic programmes, because of their size and the types of equipment in them. Some diagnostic tapes are split so that the faulty unit in the equipment is located first, and then, using another tape, the faulty component is located. Another tape division is between store and processor; these can easily be divided for fault diagnosis purposes. A disadvantage with most computer tests is that it is necessary for the computer input and output devices to be operative. If the fault is in these parts of the computer the use of diagnostic programmes becomes difficult.

SOME PROBLEMS WITH A.T.E.

It must be remembered that the purpose of A.T.E. is to improve and extend the capabilities of the technician to make his task easier and raise the efficiency of maintenance. Bearing this in mind, there are certain circumstances in which A.T.E. is a useful and sometimes necessary part of maintenance, but there are a number of problems in its use which must be overcome if its full potential is to be realized.

The first problem concerns the design of A.T.E.. Many current automatic equipments were introduced as an afterthought in an attempt to cure acute maintenance problems. These are usually not fully effective because the basic equipment design must be interfered with to accommodate A.T.E.. Connections have to be made which are time-consuming and a potential source of additional faults. Fully effective A.T.E. is therefore only likely to be achieved when it is designed as an integral part of the equipment being maintained. This does not, however, necessarily mean that it must be permanently attached to the equipment, and so making it specific to that one equipment.

A second problem concerns reliability. The user of A.T.E. must be reasonably certain that indicated faults occur in the equipment under test and not in the testing device. For this reason A.T.E. must be highly reliable, and it must have some form of built-in self-checking device if it is to reduce overall maintenance problems and not merely extend the range of equipment with which the technician has to deal. One of the problems in designing A.T.E., particularly where the complexity of A.T.E. rivals the complexity of the equipment it is designed to maintain, is therefore to achieve sufficiently high reliability.

Finally there is a human problem in the use of A.T.E.. Men are generally reluctant to accept a device which does not permit them to use their full range of knowledge and skill. They are also reluctant to accept new methods unless they can be shown to be effective and superior to existing ones. It is a clear requirement that a maintenance organization must ensure that equipment is used by the appropriate staff, and that the staff are correctly motivated, in order to achieve efficient operation. Automatic test equipment of the types outlined clearly has a major influence over the nature of the maintenance task, and hence over the selection and training of men who are to use it.

SUMMARY

When using a test instrument, technicians must always be aware of the tolerance that is allowable in the quantity being measured, of the effect of connecting the test instrument and of the accurary of the instrument. Care in handling of instruments is essential, and safety precautions must always be observed.

General test equipment is used for a wide range of measurements on a wide range of equipments; examples are multimeters and oscilloscopes. Although there are advantages in having a small number of multi-purpose instruments, there is also the possibility of an instrument not being available when it is required.

Special test equipment is used with a particular main equipment to provide for specialized measurements or to produce specialized signals, and it is essentially part of the main equipment. The amount of such equipment is best kept to a minimum.

Automatic test equipment (A.T.E.), which automates part of the maintenance task, is becoming essential for very complex equipment. Two types of A.T.E. are: audio equipment, in which instructions are given to the technician through earphones; and tape-controlled A.T.E., which connects various testing signals to the equipment in accordance with a programme on magnetic tape.

Computers lend themselves, by their nature, to automatic testing.

A.T.E. carries its own problems in respect of design and reliability. A.T.E. should be designed at the same time as the main equipment, and its reliability should be very high.

CHAPTER 11 TRAINING FOR MAINTENANCE

INTRODUCTION

However much maintenance is automated, some training for it will always be necessary. In most organizations, however, little automation has so far taken place, and traditional training is therefore a major feature of maintenance policy. Until recently little attention has been paid to the efficiency and effectiveness of training courses, but advances in knowledge of the processes of learning and an increasing national interest in training have resulted in considerable emphasis being placed on this subject in the past decade. The main advance is that training is now looked upon as a process in which both teacher and student are parts of an integral whole, the teacher playing the part of a manager, organizing and controlling his resources to suit the student, and the student acting as the product which must emerge at the end of training with all the required characteristics. In such a process, training is centred around the student, and not the teacher, in contrast to the method which has so often been adopted in the past. Situations are controlled to suit the needs of the student, his progress is frequently monitored, he works at a pace most suited to him, and he is appropriately strengthened at each stage in his training course. In this manner the widely accepted principles of learning are applied to best advantage.

A comparison between old and new learning systems is shown in Figure 11.1. Figure 11.1(a) shows a conventional system in which the strongest link is the information passed from store to student. Feedback of examination results to provide incentives to the student, and to control the information presented, is weak and sporadic. In Figure 11.1(b) the whole system is closely controlled by the feedback loop. Student responses are compared with what they should be, and the result of this comparison is used to control the presentation of information to the student.

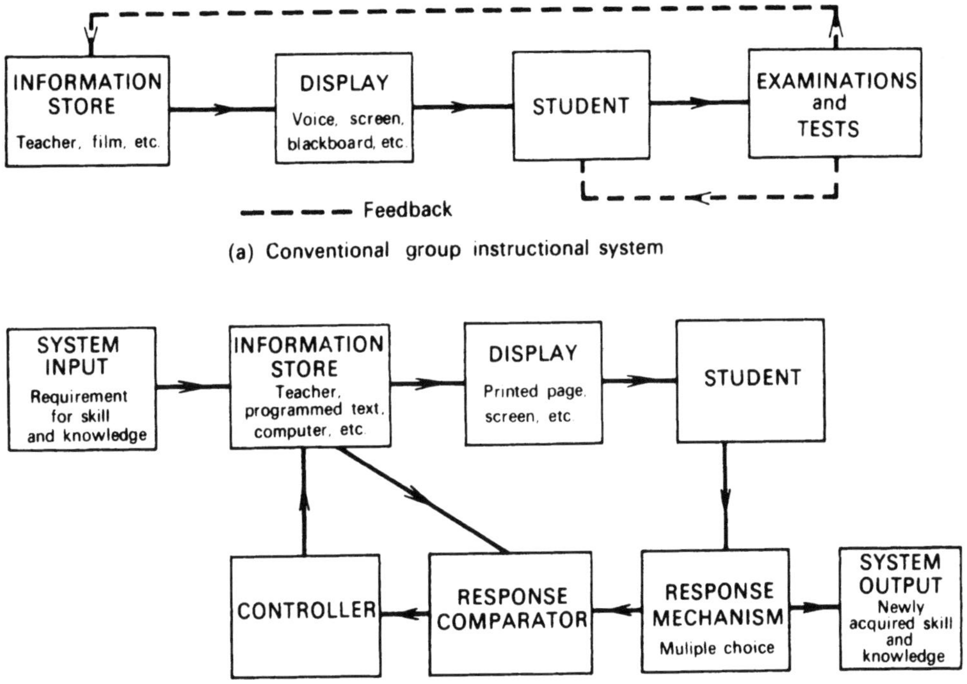

Figure 11.1 Comparison of instructional systems

Because of these advances in training technology the student is likely to meet a number of new methods and aids in a modern maintenance course. In this chapter we first discuss the content of maintenance training course and the way in which the content should be determined, and then explain some modern training aids which are likely to find increasing use in maintenance courses.

COURSE CONTENT

We have already mentioned, in Chapter 4, that the maintenance training requirement is the difference between the ability required to do the job and the ability of men selected to do it. Before deciding course content there are therefore two things to be done; one is to analyze the job which has to be done, the other is to assess the ability of men who are selected. The former is more important, as it must always be the starting point in planning training courses, and we shall therefore discuss it in more detail. The latter is governed by maintenance policy, which must be chosen to make the best compromise between selecting men of high ability, who need little training but are difficult to find and expensive to obtain, and selecting men of low ability who require more training but are easier to find and less expensive. The selection of men at the required level is the province of the skilled personnel selection officer using selection tests, reports and interviews; this topic will therefore not be dealt with in this book.

JOB ANALYSIS

There has been a tendency in the past to extend and intensify maintenance courses to cover new electronic design techniques and increases in complexity. This is acceptable where courses are short and inexpensive, but in recent years all forms of electronic training have tended to become long and intense. The stage has been reached where it is essential to minimize the increases in length and complexity, and this is being done by looking critically at what is taught to ensure that it is relevant to the aims of the course. Only in this way are increases being reduced and courses made more efficient. The aims of a course arise from the job requirement, so that attention has been focussed on the all-important work of job analysis.

The object of job analysis is to determine the knowledge, skills and attitudes which are essential for efficient performance of the jobs. This is done by detailed examination, not only of what tasks a man has to do, but also what equipment is operated upon and what tools or aids he uses. Such an analysis ensures that only relevant material is included in training courses. Job analysis has already led to drastic changes in course content, and in many cases it has also led to changes in the job itself, e.g. by revealing the presence of unnecessary tasks or of inadequate aids.

One of the difficulties in determining course content is that a man is usually being trained to do a variety of jobs, and gain a number of other attributes, in a single course. Not only does he learn the job itself, but also, for example, he is often expected to improve his general education, keep physically fit and learn about organization; thus a course can consist of a wide variety of topics, not all directly related to the job. In a well-planned course the aims which each part of the syllabus set out to satisfy should be clear, and apparent to both teacher and student.

There are a number of ways of carrying out a task analysis, and they are usually used in a combination to obtain the best overall result. All analysis must start with an overall assessment of the job; this includes examination of the policy, equipment, locations and the duties involved in the job. This is backed either by direct observation of men carrying out tasks, or controlled experiment, or comparison with analyses which have been already completed. The first of these requires that the job is already being done by someone; it also involves time-consuming observation although this can be reduced by work sampling, and the careful selection of tasks to be observed. Observation of tasks also involves the problem of having to determine from a man's actions what knowledge and skill is required to carry out the actions. Controlled experiment is more likely to provide reliable information about the relationships between certain topics and job performance, but it is even more time-consuming than observation. The experimenter has always the dilemma of either so closely controlling his experiment that he makes it unrealistic, or making it realistic and losing close control. The third technique of comparison is by far the easiest but, at present, the least precise because there is a lack of methods of comparison which reflect training requirements. The method therefore leans heavily on the opinion and judgement of a small number of experts.

In normal work a technician is unlikely to be involved in job analysis, but in his training course he will experience the results of such analyses. For this reason we will briefly describe some of the findings of job analyses which are applicable to maintenance training courses.

THE EFFECT OF AIDS TO MAINTENANCE

For men of a particular ability level the greater the amount of aid which is given to a job the less is the training requirement. It is not surprising, therefore, that job analyses have shown aids to be the greatest single factor which controls course content. A critical look at aids to maintenance has also resulted in improvements being made to them; the improved diagrams described in Chapter 8, and the fault finding guides based on them which are described in Chapter 9, are examples of improvements which have been made as a result of job analysis.

THE ROLE OF ELECTRONIC THEORY

Theoretical instruction has traditionally occupied a major part of all electronic maintenance courses, and it is not surprising that a considerable amount of work has been done in order to determine its importance in job performance. The results of this work have shown that, for the non-routine maintenance task, a thorough grounding in theoretical electronics is not in itself a guarantee of good performance. This should not be taken as an indication that theoretical training is not required, but that its importance lies in training aims other than carrying out a specific job; for example, there are considerations of educational requirements, training for design tasks, and career considerations.

It is also true that theoretical knowledge is necessary for difficult fault diagnosis tasks such as locating a fault without the use of aids, or locating faults due to poor design. These tasks should be exceptions to normal fault diagnostic work, and allowance should be made for them in the organization of maintenance, by having available members of the staff who are trained to cope with these exceptions.

METHODS OF FAULT LOCATION

The task demanded of a person locating faults in electronic equipment is to search for a defective part among a large number of integrated parts which together fulfil an equipment function; methods of fault location were discussed fully in Chapter 5. Research has shown that men cannot be expected to decide on the best method to use without training; it has also been shown that, in practice, equipment and aids must be in a form which encourages the use of the best methods. Not only must methods of location be taught, but they must also be exemplified and emphasized in practical exercises, and be supported by aids which encourage the use of the best method of location.

In order to apply efficient methods of fault location it is necessary to have functional knowledge of an equipment, which consists essentially of an understanding of the inter-relationships among the component parts; research has supported this requirement by correlating tests of practical performance with tests of functional knowledge. The way in which diagrams are drawn to illustrate functional information in the best possible way was explained in Chapter 8.

TRAINING METHODS

Having discussed the methods of determining training course content and some of the important topics in a maintenance training course, we will now explain some modern methods of presenting information to a student. Before doing this, brief mention will be made of the best order in which to present information. In the past it has always been considered that training for a job should gradually build up by commencing with fundamentals and expanding these to show how they are applied in the circumstances of the job itself. This is really the way in which a designer builds up to his concept of an equipment. In training for a job it is much more logical, and effective, if training is given in an order which corresponds to the stages of the job. In fault diagnosis, for instance, the task commences with a defect report; good training for this task would also start with the defect report, then continue with the functional check and fault location. Aids to the task and fundamental principles are only introduced as and when they are required. In this manner a student is more easily able to see the reasons for each aspect of his training and thus will be more effective in carrying out his work. Many students are discouraged in their studies by commencing a course with a large amount of apparently irrelevant material; in a modern course he would commence with something which is both relevant and interesting.

PROGRAMMED LEARNING

The biggest advance of recent years in education and training methods has been the development of programmed learning. Programmed learning is the presentation of information to individuals in a standardized fashion, and in such a manner as to take advantage of the incentive derived from the immediate feedback of test results. Course material is presented as a sequence of items consisting of information and instructions. This sequence of items is called a programme, and the devices used to present the programme, obtain responses and provide feedback are called **teaching machines.** Television, tape recorders and film projectors do not at present satisfy this definition, and they must continue to be looked upon as teaching or training aids, and not teaching machines.

Programmed learning emphasizes the role of the student in the learning process; he learns for himself, at his own rate, and is appropriately reinforced and well motivated. These are the major requirements of an efficient learning system and it is not surprising therefore that the use of programmes and teaching machines is expanding rapidly.

The programme is more important than the machine, and considerable time and effort is necessary to prepare good programmes; in fact the difficulty of writing good programmes is the major factor in hindering the expansion of programmed learning. There are a number of programming methods in use and more are likely to be introduced as more becomes known about the relationship between programming methods and different teaching or training requirements. For instance, the type of programme selected for teaching typewriting, which is a manual skill, would probably be different from that selected for teaching Ohm's Law. Although a number of programming methods may be encountered, most of them stem from the two original and distinct methods called **linear** and **branching.** These two methods of programme construction will be explained in detail.

Following extensive experimental work on animal behaviour, Professor Skinner of Harvard University developed a theory of learning and applied it to human beings in the form of **linear** programmes. An example of a linear programme is shown in Table 11.1.

TABLE 11.1: EXTRACT FROM A LINEAR PROGRAMME

Frame	Stimulus	Response
1	Metals expand when heated. Copper is a metal and will .... when heated.	
2	When heated iron will .... less than copper.	Expand
3	Thus different metals will expand by .... amounts when heated.	Expand
4	The increase in length is proportional to the .... in temperature.	Different
5	A copper bar expands by 0.2 cm when the temperature rises by 100°C. It will expand .... if the temperature is increased by 200°C.	Increase (Rise)
		0.4 cm

The material to be learned is split into very small units, called frames, containing between 10 and 20 words. Each frame contains a question (stimulus) of the completion type to which the student must provide an answer (response). The stimuli are carefully prepared to ensure a high success rate and no wrong responses are shown for the user's consideration; he simply checks his response against the correct one which is revealed to him when he moves to the next frame in the programme. All students take the same route through the programme, giving all (or nearly all) correct responses and thus receiving continuous reinforcement. Table 11.1 can be used to show how a linear programme is used. Frames 2 onwards are covered while frame 1 is read and the response made. Frame 2 is then exposed showing the next stimulus and the correct response to Frame 1. This sequence is continued throughout the programme. It should be remembered that Table 11.1 is only a small extract from a very long programme directed at students of a particular ability level, not necessarily the same as that of the reader. Linear programmes give continuous reinforcement, individual tuition and allow the student to work at his own rate. A large number of linear programmes are commercially available; Basic Electronics, Basic Transistors, Electronic Trouble Shooting and Mathematics for Electronics, are examples of programmes which are in use at present.

Linear teaching machines require simply provision for revealing the frames of the programme sequentially and allowing the user to respond to the stimulus before a correct response is revealed to him. Most common are devices using a roll of paper containing the programme, a window to reveal stimuli and correct responses, and an aperture through which to make written responses. Book versions of linear programmes usually use a piece of card to cover responses.

Branching (sometimes called **intrinsic**) programmes were developed in America by Dr. Norman Crowder; they were the result of a series of empirical investigations rather than fundamental research into the learning process but, nevertheless, have proved to be the basis of a powerful learning method. In the branching programme a student's response is used to determine and control the order of presentation of the programmed information. Frames are presented to the student and his understanding of the information in them is tested using a question with multiple choice answers. If a correct answer is given the student goes on to the next information frame. If a wrong answer is given the student is directed through remedial frames, so that students of different ability take different routes through the programme. The information frames of a branching programme are larger than those required for a linear programme, though the writing of remedial frames makes the amount of programming which is necessary almost as great as that required for linear programmes.

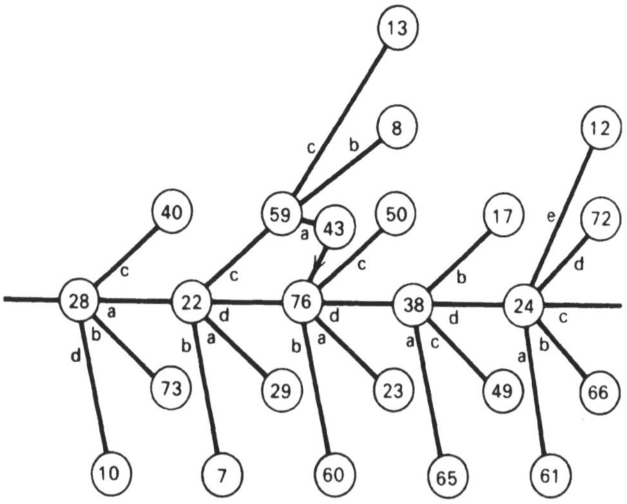

Figure 11.2 Extract from a programme bubble diagram

A small extract from a branching programme

would be meaningless, so, to illustrate their construction an extract from a programme 'bubble diagram' is shown at Figure 11.2. Each circle represents a frame. The main information frames lie along the centre of the diagram and the remedial frames on either side. The remedial frames either pose a further question with multiple choice answers (e.g. 59) or, after giving additional information to the student, direct him to attempt again the question he has answered wrongly (e.g. 40). The example shown was put into book form so that each frame is a page. The pages are arranged so that question and answers do not appear on adjacent pages. The numbers in the circles represents page numbers and the letters represent the answers to the questions. Thus a route through the portion of programme shown might be: Page 28, answer b, 73, 28, answer a, 22, answer c, 59, answer a, 43, 76 answer d, 38, answer d, 24 answer c.

Branching programmes are usually written so that only a small number of errors should be made. The student is therefore reinforced by a high degree of success, he has individual tuition and works at his own rate, although, unlike the linear programme, different sequences of frames may be used by different students. Branching programmes are available for such topics as: Fundamentals of Electricity, Introduction to Electronics and Logical Electronic Trouble Shooting.

Branching programmes do not require a student to make a response other than by selecting an answer from a list of possible answers. The process is thus easily automated and in most machines a student is required only to read the frames and press buttons. There is a wide variety of machines available, unfortunately not all compatible with regard to the use of programmes. Many of them use film as the information store with an optical display system and mechanical means of moving the film to the required frames. Like linear programmes branching programmes can be presented in book form and, in this form, they are known as **scrambled books**. Each response to a question is on one page of the book and the user selects an answer and moves to the page indicated by the answer; he does not read the pages in the order in which they are bound – hence the name scrambled book. The amount of control over the learning process is less with a book than a machine; it is difficult to keep records of the route used and incorrect responses made, and the user can cheat. For these reasons a machine has been found to be better than a book for presenting a programme.

Another device which has been found to be very effective is the adaptive teaching machine. In this the blocks of information presented to the student are made easier or harder according to the responses given. Factors such as time allowed, and amount of information given, can be used to vary the difficulty of a task or question.

COMPUTERS

The storage capacity and speed of operation of a computer makes it a tool which will be increasingly used for training, both as part of the learning method and as an aid to efficient administration.

When used as part of a learning method, computers can extend the role of the teaching machine by allowing greater flexibility of the basic variables and an increase in the number of variables involved. For instance, the multiple choice questions of a branching programme are normally limited to three or four answers, but a computer could allow this to be increased considerably. The next frame in a branching programme is usually determined only by the response given to the current frame, but a computer would enable the choice to be based upon as large a number of previous responses as was considered relevant. An extension of variables could allow programme sequences to be governed not only by analysis of response but also by other abilities and personality traits.

A computer clearly cannot be provided for each individual, and it could only be justified if it provided a large number of terminals or response stations, which could be remote from the centrally situated computer. One forecast is that 500 to 1,000 students will ultimately use one computer simultaneously, but at present the 32 terminals of one currently operating system is a more typical situation.

One of the most important features of any system is the terminal or response station, which serves both for information display and for student response. It is crucial because it is the junction at which student and machine come into contact, and any limitation here may restrict the usefulness of the entire system; similarly improvement in the response station may increase the utilization of the capabilities of the computer. The most common type of response station is simply a computer-controlled electric typewriter. The computer prints information and the questions, and the student replies by typing his responses on the same machine. Other devices may also be used, such as random access slide projectors, television, and tape recorders. One device developed specially for such a system is a 'light pen' which enables a student to write his response directly on the face of an oscilloscope as he would write with pen on paper.

The computer is clearly a powerful tool which can considerably improve the quality and efficiency of training. Its chief disadvantages are high cost, lack of mobility and the inability of teachers to use its potential to the full, bearing in mind current knowledge of the learning process. The computer could be used for research into individual differences by using such detailed information as response times, incorrect responses given and total learning time.

For administration the computer also has considerable potential. A vast amount of time is spent on such things as preparing timetables, marking question papers and analyzing examination results; all of these are tasks which could be done more efficiently by computer and so release teachers for more important aspects of their job. Again central computers could provide a service for a large number of training establishments.

TRAINING AIDS

Teaching machines and computers can be used as methods of training in their own right. There are a number of other devices used in modern training courses which are useful and effective aids to training but cannot normally be considered as methods of training; such aids are tape recorders, television, films and overhead projectors.

Audio tape is a well established training aid, although it generally has greater application to language training than to maintenance training. Video tape has much wider application with its ability to record complete lectures and repeat them at any convenient time or place.

Closed circuit television (CCTV) is a method of communication between teacher and student which is being used increasingly. Its main advantages lie in its ability to demonstrate equipment which for reasons of small size, difficulty of access, or safety could otherwise by seen by very few students, and its ability to deal with large class sizes, or more than one class. The disadvantages of CCTV are high cost, lack of feedback from students, and the need for experts to operate and produce the televised lessons. The qualities of a good instructor are not the same as those required by a good television producer or operator, so care has to be taken that the advantages offered by television are not offset by production of poor quality. Probably the best CCTV system is one which incorporates a video tape recorder to play back lessons and demonstrations recorded by instructors who are experts in the use of the medium. The facilities of CCTV can also be increased by making provision for a telecine input to the system, enabling film, slides and caption cards to be displayed, and by arranging for the reception of broadcast programmes. A typical arrangement of all these facilities might be similar to that shown in Figure 11.3. Audio links, which could provide a degree of communication between demonstrator and student via a control console, might also be included in the system.

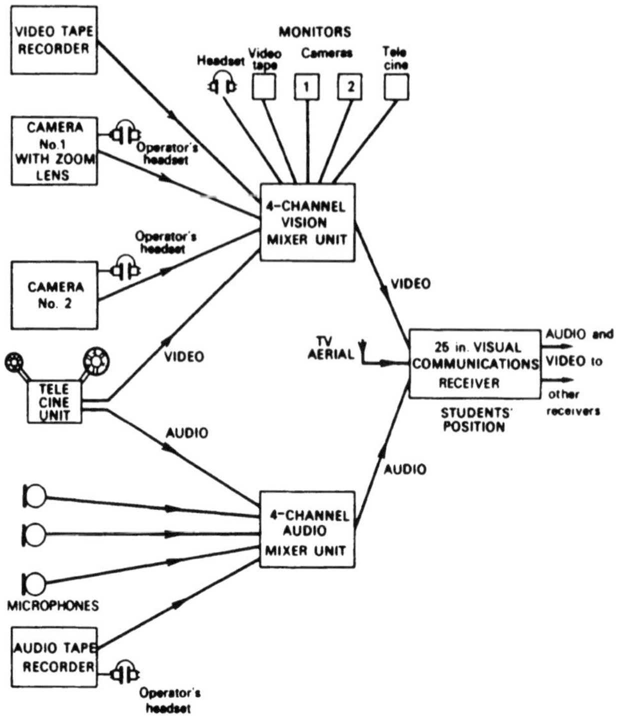

Figure 11.3 A closed circuit television system

Conventional films and slide projectors are commonplace in training establishments, and more recently the overhead projector has also been found to have wide application. In this device a transparency is placed on a horizontal illuminated plate and the image is projected via a 90° prism on to a screen.. The user can operate it while still facing his class, or sitting at his desk; transparencies can easily be changed; details of a transparency can easily be pointed out; the diagram can if necessary, be added to, by using a special pen, and transparencies are easily made. The overhead projector has possibilities for many other demonstrations; for example, it has already been used for demonstrating the dynamic behaviour of fluids by using water circulated through an enclosed perspex channel which is located over the light table.

SUMMARY

The importance of the correct training of technicians to carry out maintenance has been stressed. In designing a suitable training course the following factors have to be considered.

1. The content of the course must be specifically decided by careful analysis of the jobs that the trained product is expected to undertake, and realization of the ability of the students themselves.

2. The amount of theory which is necessary should be critically examined.

3. Logical methods of fault diagnosis should be taught.

Some modern methods of presenting information to students have been described, such as programmed learning and using computers. Some modern training aids have also been described, with their advantages and disadvantages, such as audio tape and closed circuit television.

CHAPTER 12 MAINTENANCE RESEARCH

INTRODUCTION

In this chapter we shall describe some of the work which has been carried out with the primary object of improving understanding of the non-routine maintenance task and, at the same time, determining the training which is required for it. This is the task analysis approach to training which was explained in Chapter 11. No attempt will be made to describe research which has been undertaken to improve the maintenance factors concerned with hardware e.g. reliability or automatic test equipment.

A lot of maintenance research has been done in both Great Britain and America, particularly in the Armed Services, and it would be impossible to summarize it all in a book of this length. A few examples of the work have been selected as typical and these will be explained in some detail. It must be remembered that they are only examples, and the current state of our knowledge of the task, as reported in this book, is based on the accumulation of evidence over a long period and from a large number of sources.

Since more maintenance research has been done in America than anywhere else, we have selected three American projects from a research programme organized by the Training Methods Division of the George Washington University, Human Resources Research Office (HumRRO), operating under contract with the Department of the Army.

GUIDED MISSILE MAINTENANCE

The object of the first project was to study methods by which the training requirements for guided missile technicians could be derived from an examination of the work done in the field; emphasis was placed on the knowledge and skills required for the work.

A tracking radar was used as the research equipment and the analysis was carried out under four headings, viz. system analysis, task analysis, knowledge and skills analysis, and determination of training objectives. The results obtained from the analysis were used to prepare a maintenance training course; this course was given to a group of students and the results of an end-of-course practical test were used to compare their performances with two groups of conventionally trained technicians.

SYSTEM AND TASK ANALYSES

Under the heading of system analysis information about the necessary maintenance work was collected by studying the design of the radar equipment, the maintenance policy and the aids to maintenance. It was also found necessary to collect additional information, not normally documented, on the nature and frequency of maintenance tasks, and the availability of aids and assistance; this was done by issuing questionnaires to all technicians working on the equipment, by examining maintenance records from both the Service user and the contractor, by carrying out interviews and observations, and by obtaining information from a number of other minor sources. Altogether this data provided a comprehensive account of radar maintenance work.

The next stage was to analyze the work into tasks. This was done by identifying cues and responses, a method adopted in numerous other research projects. Cues, or stimuli, consist of indications of the equipment's condition, such as meter readings, oscilloscope displays and other test equipment outputs. Responses are the actions taken by a technician following the recognition of a cue; examples are operating a switch, adjusting a control and changing a valve. In a fault location procedure a cue given by a test point measurement should result in a response, which is either a further check to narrow down the faulty area or the identification of a faulty component. Such an analysis ensures that any sequence of checks is systematic, but, to ensure that it is also efficient, it is necessary that the equipment and aids are in a form that encourages the use of the best location methods. It must be possible to make conclusive tests at the points required by the method, and a means of making the tests and interpreting the results must be available.

In the case of routine maintenance, the task analysis proved to be relatively simple because the schedules which were provided for the equipment gave the necessary information; for non-routine maintenance the task analysis proved more difficult because each component in each possible failure mode could provide a unique set of cues and responses i.e. could require a unique sequence of checks for positive location. For example, a short-circuited resistor could provide a set of fault symptoms different from the symptoms obtained with the same resistor on open circuit.

Since all fault conditions could not be investigated a sample was selected, and every effort was made to ensure that the sample incorporated as many as possible of the faults which would be experienced in practice. The half-split rule, taking into account the ease of making a check, was used as the basic fault-location method, though it was frequently found necessary to use, in addition, alternative methods where application of the basic one proved too difficult. For example if fault symptoms are reflected back through a number of units, a method based on unit replacement might be more economical. An analysis of the steps taken to locate each fault in the sample was done theoretically using equipment diagrams, but where there was any doubt about a route, it was checked by practical tests on the equipment.

DERIVATION OF KNOWLEDGE AND SKILL REQUIREMENTS

Having completed the task analysis it was necessary to identify the knowledge and skills required to enable a technician to detect the cues and make the appropriate response involved in each task. This was the part of the project relating to knowledge and skill analysis. It was carried out under the heading of **knowledge**; which included procedures, factual data and theory; **motor skills**, such as repair and alignment; and **mental skills** such as deriving required information from more fundamental information, making logical inferences, making comparisons and carrying out calculations.

DETERMINATION OF TRAINING OBJECTIVES

The final stage was to determine training objectives from the findings of the analysis. This was accomplished by taking the full range of job requirements in terms of knowledge and skills, and subtracting those which were obtainable from other sources than training, such as aids to maintenance, formal on-the-job training and experience; the residue formed the basis for the training objectives.

A training course was developed to satisfy the training objectives. Its content stemmed directly from the findings of the task analysis and the training methods and media were also, necessarily, changed to accommodate the changes in content. The final course content is summarized in Table 12.1, which shows the emphasis placed on topics directly related to fault location.

TABLE 12.1: SUMMARY OF EXPERIMENTAL COURSE CONTENT

Item	Type of Content	Content
1	Procedures	a. **Specific procedures.** Includes preventive maintenance schedules, component testing, modifications and other procedures covered in available aids to maintenance.
		b. **General procedures.** Operation of the radar, operation of general test equipment, removal and replacement of components, repair, adjustment, items of preventive maintenance, safety precautions, general checking procedures required during non-routine maintenance.
2	Technical data	a. **Physical location and identification** e.g. of controls, test points, components.
		b. **Part characteristics** E.g. the normal and likely fault conditions of components, corrosion, charring, noise.
		c. **Normal outputs** Descriptions of all the equipment outputs including test points, tolerances and ranges of variation of outputs where appropriate.
		d. **Signal flow** Information on the functional structure of the equipment.
		e. **Failure data** Reliability, relative likelihood of different types of failure, and likely recurrence of rectified faults.
		f. **Symptom-cause information** The equipment's function in functional terms.

Item	Type of Content	Content
3	Theory	a. **Fundamental electronics** This was confined to the characteristics of voltage, current, resistance, power, capacitance and reactance, and resistors in series, parallel and series-parallel combinations. This is information required for component and wiring tests. b. **Circuit operation** Confined to types of stage rather than specific stages; covering d.c. voltage conditions, qualitative explanation of the circuits' operation, effect of component failure on the output, and components whose failure could only be detected by substitution.
4	Motor skills	Primarily activities requiring soldering, part removal or delicate adjustment.
5	Mental skills	The best methods of fault detection and location. The half-split method and rules derived from it.

The method of training used what is termed a 'functional context' approach in which training content is made more meaningful by relating it closely to job performance. In such training the sequence of instruction is whole-to-part. It begins with the overall system and progresses via equipments and removable units to components. A concrete-to-abstract sequence is also used wherever possible so that observable aspects of maintenance, e.g. control settings and display characteristics, are taught first, to enable more abstract functions, e.g. detection and demodulation of a signal, to be derived from a practical framework. Other characteristics of the functional context approach are that individual items of content are taught as and when required in the job; for example, test-equipment and tools are introduced only in the context in which they are used in the actual work; and nomenclature is introduced only after the concept, law, procedure or equipment to which it relates has been dealt with.

To accommodate these changes in training content and method it was also found necessary to change other aspects of training. The main changes were that diagrams were prepared which were compatible with the whole-to-part sequence of instruction (this is a characteristic of functional diagrams), and increased emphasis was placed on practical work. Changes in performance tests were made that also reflected the increased emphasis on the practical aspects of maintenance.

In its final form the course was 22 weeks (836 hours) long with 440 hours of classroom instruction and 396 hours devoted to equipment. The conventional course also lasted 22 weeks.

EXPERIMENTAL EVALUATION

The means used to evaluate the effectiveness of the training programme was to give the course to a group of 12 students, testing them at the end of training and compare their performance with the performance, in the same test, of a group of 12 technicians who had just completed the conventional training course, and also of a group of 12 technicians who had 1 year's field experience in maintaining the equipment.

The performance test consisted of 92 tasks selected from a sample of 300 tasks collected in a field survey. Only 23 of the 92 tasks were given to each man, but this took approximately 11 days of testing. Most of the tasks involved all three aspects of non-routine maintenance, i.e. fault detection, location and repair; they were administered under conditions which, as closely as possible, simulated those existing on the job. Proficiency was measured primarily by the speed with which the tasks were accurately and safely completed, and the equipment returned to normal operating tolerances. A summary of the tasks used in the test is shown in Table 12.2; it can be seen that all the important aspects of maintenance were covered, with particular emphasis on fault-location.

TABLE 12.2: SUMMARY OF TASKS USED IN THE TEST

Area	Category	Description	Number of tasks
Non-routine maintenance carried out in the workshop.	Component fault finding.	Acceptance testing, location and repair of defective components using the ORD6 test equipment.	31
	ORD6 fault finding.	Location and repair of defective components in the ORD6 test equipment, detected during component fault finding.	10
Non-routine maintenance of the system.	From symptoms.	Location and repair of faults detected during the normal operation of the equipment.	28
	From checks.	Location and repair of faults detected during operational checks of the equipment.	9
On-site maintenance	Component fault finding.	Repair of known faulty components in the system.	6
	Replacement.	Replacement of components and parts according to a routine maintenance schedule.	8
		TOTAL	92

Analysis of the test results showed that the experimental group did significantly better than those who had recently finished the conventional course, both in the overall test and in the parts of the test dealing with non-routine maintenance of the radar system and component fault-finding. The differences obtained in routine repair tests were not significant. The experimental students proved to be almost equal to the experienced group in component fault-finding but worse in other categories. No statistical comparison was made between the experimental and experienced groups.

On the basis of the results it was concluded that the methods used in the study were capable of identifying appropriate work and training requirements, and that greater emphasis needed to be given, during maintenance training courses, to the type of content used in the experimental courses. This book has explained in detail the major new items of experimental course content and the important training methods used in the project.

RADIO EQUIPMENT MAINTENANCE

The object of the second project which we will summarize was to improve technician training for the maintenance of ground radio communications equipment. A trade which covered several equipments, and which was judged to be typical of a range of maintenance trades, was chosen, so that the results obtained would have a wide application. The project consisted of carrying out a field survey of the technicians' work, preparing a proficiency test, developing and evaluating a revised training course, and carrying out a follow-up evaluation after students of the experimental course had gained field experience in maintaining the equipment.

A number of methods of collecting data were used for the field survey and about 1,000 maintenance technicians, and their supervisors, were involved. The results suggested that more emphasis in training should be placed on the recognition and correction of common faults in the most unreliable equipment, that training should be directed more specifically towards maintenance, that an optimum general fault location method should be taught and that the principles of functional context training, as described on page 76 of this book should be adopted. The field study also provided data for the preparation of the proficiency tests. In all, seven tests were used. Four were practical: fault location, test equipment, repair and alignment. Three were written: manuals, schematics and achievement. The fault location test lasted approximately four hours and consisted of four problems per student, taken from a sample of eight, which covered five equipments. The test equipment test required the subject to make tests and interpret results. The three written tests were all of the multiple choice type; the manuals and schematics tests required subjects to locate and understand the information provided; the achievement test consisted of fault location and repair questions.

TABLE 12.3: KNOWLEDGE AND SKILL REQUIREMENTS

Item	Type of Work	Knowledge or Skill
1	Fault Location.	**a**. Knowledge of, or ability to acquire information about, the operational characteristics and capabilities of equipment as a basis for the recognition of fault symptoms. **b**. Ability to recognize visual fault symptoms. **c** Knowledge of, or ability to acquire information about, the functional relationships among circuits, components, and subassemblies, as they are revealed through operational or test equipment indications. **d**. Knowledge of common symptom-cause relationships. **e** Ability to plan and carry out a sequence of fault-location checks according to given optimum rules. **f** Ability to use schematic diagrams for the identification of signal paths, current paths, test points, stages, and subassemblies: ability to relate diagrams to equipment. **g**. Ability to select and set up appropriate test equipment for particular applications; ability to interpret test equipment indications. **h**. Ability to make the best use of technical publications. **j** Ability to use sections of standard alignment procedures as an aid to fault location.
2	Fault rectification	Ability to select and use the correct hand tools for repair, particularly the use of the soldering iron in miniaturized equipment. Knowledge of, and proper attitude towards, the care of tools.
3	Alignment.	Ability to use written alignment procedures and specifications; ability to select and use appropriate test equipment.
4	Administration.	Knowledge of the regulations and administrative procedures governing the work.

TABLE 12.4: COMPARISON OF STANDARD AND EXPERIMENTAL COURSES

Topic	Hours of Instruction	
	Standard	Experimental
Introduction to the school.	6	6
Fundamentals of radio maintenance		
Basic electronics	90	32
Test equipment	16	13
Characteristics, preventive maintenance, operation and alignment of the training equipment.	14	29
Fault location and schematic diagrams.	33	78
Repair	7	15
Examination	12	5
Military communication equipment		
Introduction to the equipment; principles of FM.	10	16
Characteristics, operation and preventive maintenance	71	68
Purpose and function of the major stages and schematic diagram analysis	187	157
Alignment and fault location	177	165
Repair	0	13
Examination	51	52
Practical exercises under normal working conditions.	62	87
TOTAL ACADEMIC HOURS	736	736
TOTAL WEEKS	22	22

The experimental course content is summarized in Tables 12.3 and 12.4. Table 12.3 summarizes the knowledge and skill requirements which were derived from the field survey, and on which the course content was based. By comparison with Table 12.1 it can be seen that both projects arrived at similar conclusions with regard to essential knowledge and skill requirements. Table 12.4 summarizes the time allocated to the various course topics and compares these times for the standard and experimental courses. The overall course length was intentionally made the same, but the distribution of time between the two courses is considerably different; in particular, under the fundamentals heading, the allocation to basic electronics in the experimental course is only about one-third that of the standard course, and the allocation to fault location and schematic diagram training (recognition of functional structure and relating diagrams to hardware) is more than double. Nine equipments were included under the heading Military Communication Equipment and, although the total times under this heading are approximately the same for the two courses, the distribution among the equipments was quite different because the allocation of time in the experimental course was closely matched to the work requirement.

The methods of training used were governed by the requirements to adhere closely to the methods of normal field work, to implement the principles of functional context training and to give the maximum possible amount of realistic practice in fault location.

EVALUATION OF THE EXPERIMENTAL COURSE

To test the validity of the methods used to develop the experimental course, 86 men were trained in it and their performance in the proficiency test was compared with a matched group of 86 men trained using the conventional course. The results obtained in the seven sections of the test are summarized in Table 12.5. It can be seen that the experimental group was superior on four of the seven tests, viz. fault location, repair, test equipment and achievement; but neither group was superior on the manuals, schematics and alignment tests. It is particularly interesting to note the superior performance of the experimental group in the fault location test; detailed analysis of performance in this test showed that the result could occur by chance less than once in a thousand tests. For all 8 problems more of the experimental than the standard subjects obtained correct solutions, in spite of the fact that three of the problems were on equipments which were not selected for detailed treatment in the experimental course.

The aim of the third HumRRO project to be summarized was, therefore, the prediction of training requirements for new equipments. It was undertaken in three phases. The object of the first phase was to develop and test methods for analyzing electronic equipment which defined a set of knowledges and skills for operating and maintaining it. The second phase investigated the possibility of using mock-up equipment, in place of actual equipment, for training; this is a necessary requirement when the actual equipment cannot be made available for training purposes. This phase also investigated the ability of students to use the knowledge and skill which were gained in training to maintain different equipment. The third phase was the implementation and checking of the maintenance concept using a new equipment.

For the first phase an equipment of sufficient complexity to present maintenance problems similar to those predicted for future systems was used. An analysis of maintenance work was carried out using the cue-response method described on page 75, but based only on the type of information which had been available before production of the equipment. The method was tested by comparing the results which were obtained following training based on the work analysis with results obtained following traditional training. Twenty technicians were trained in 400 hours, and then tested using a nine-day individual performance test on the equipment. Seventeen technicians with 1,000 hours of standard training were given the same test. Despite the facts that the experimental training time was less than half the standard training time, and that all identified factors affecting group performance e.g. training methods were made equivalent or favouring the standard student, there were no practical differences in proficiency between the experimental and the conventionally trained groups. The performance level of both groups was

TABLE 12.5: SUMMARY OF RESULTS

	Practical tests	Written tests
Experimental course subjects were significantly superior in:	Fault location. Test equipment. Repair.	Achievement.
Experimental and standard course subjects did not differ significantly in:	Alignment.	Manuals. Schematics.

FORECASTING WORK REQUIREMENTS

The two projects already described leaned heavily on a detailed examination of the work being done on equipment during its useful life. This means that the techniques may not be applicable to the period immediately after the introduction of a new equipment, a period which, because of the lack of information on reliability and the lack of technicians experienced in its maintenance, is likely to prove a difficult one. Clearly it is necessary to develop procedures which enable the work and training requirements of new equipment to be forecast so that training and production may proceed together and, when the equipment is introduced to operational use, trained technicians are available to service it.

about 50 per cent of the total possible score.

To evaluate the second phase, six students were trained with the same course content but with specially designed mock-up equipment and a 10 per cent longer training course. In the same test they achieved an average proficiency level of 70 per cent. Another group of 5 students was trained in the same way but equipment was used for 80 per cent of the course and another equipment, on which they were tested, was used for the remainder. The result was a mean score of 70 per cent.

In phase three of the project the techniques developed in the first two phases were applied to improve a ground radar equipment training course. The course was a combination of traditional and new types of content and methods, with the new portion directed primarily at fault location to unit level and overall equipment functioning. The training course has been in use for a considerable length of time and has received favourable comment from both supervisors and students.

The conclusions and implications arising from this extensive research project, which even so are not fully mentioned above, can be summarized as follows:

a. Training based on the type of analysis used results in effective maintenance work and decreased training time.

b. The method can be successfully used to predict the training requirements of new equipment.

c. Task analysis provides the basis for an effective fault location method.

d. Maintenance aids can be improved to facilitate work on an equipment.

e. The ability to use the method of analysis can be taught to experienced, competent technicians and used by them to develop training content.

f. Maintenance proficiency may be increased, and the cost of training equipment reduced, by the use of simulators as fault location training devices.

g. The method of training enables technicians to work on a range of equipments, similar to those used in training, with a minimum amount of additional specific training.

SUMMARY

It is maintenance research of the type summarized above that has led to a much better understanding of the nature of the technician's work and the requirements for its efficient execution. Research is, as yet, incomplete, but it is possible to make the following statements which are based on a variety of experiments and sources; they have already been mentioned in Chapter 11, but since they form the basis on which the philosophy of maintenance explained in this book is based, they are repeated here.

a. Maintenance aids, more than anything else, determine the nature of maintenance work and hence the training required for it. It is possible to improve performance considerably by relating aids more closely to work requirements than has been done in the past. Functional diagrams and fault finding guides based on functional structure are examples of aids which are designed to provide the critical information for fault location. Both have been evaluated in research tasks which have resulted in favourable conclusions.

b. Training course content is most effectively obtained by analyzing work being done on the equipment and deriving knowledge and skill requirements from the analysis. The cue-response technique is an example of a proven method of analysis.

c. It should not be assumed that a thorough grounding in theoretical electronics is, in itself, a guarantee of good fault location performance.

d. Optimum fault finding procedures can be taught, but these are unlikely to be effective unless the equipment and the aids to maintenance are in a form which enables them to be used.

Modern Electronic Maintenance Principles

INDEX

INDEX

Printed in France by Amazon
Brétigny-sur-Orge, FR

51786490R00056